It's another great book from CGP...

OK, so GCSE Maths can be seriously challenging — and the latest courses are tougher than ever. To do well, you'll need plenty of practice at answering the type of questions you'll face in the final exams.

And you guessed it... this brilliant CGP book is brimming with realistic exam-style questions, all up-to-date for the latest GCSE requirements!

We've also included fully-worked answers, so if you drop any marks, it's easy to find out exactly where you went wrong.

CGP — still the best! ☺

Our sole aim here at CGP is to produce the highest quality books — carefully written, immaculately presented and dangerously close to being funny.

Then we work our socks off to get them out to you
— at the cheapest possible prices.

Contents

✓ Use the tick boxes to check off the topics you've completed.

Section Five — Shapes and Area

Section Six — Angles and Geometry

Section Seven — Probability and Statistics

Practice Papers

How to get answers for the Practice Papers
Worked solutions to all three practice papers are available online for you to download or print.
Go to **www.cgpbooks.co.uk/gcsemathsanswers** to get hold of them.

Published by CGP

Editors:
Rob Harrison, Shaun Harrogate, Sarah Oxley, David Ryan, Caley Simpson, Ruth Wilbourne.

Contributors:
Alastair Duncombe.

With thanks to Jane Appleton and Alison Palin for the proofreading.

Clipart from Corel®
Printed by Elanders Ltd, Newcastle upon Tyne

Based on the classic CGP style created by Richard Parsons.

How to Use This Book

- Hold the book <u>upright</u>, approximately <u>50 cm</u> from your face, ensuring that the text looks like <u>this</u>, not ͛ᴉɥʇ. Alternatively, place the book on a <u>horizontal</u> surface (e.g. a table or desk) and sit adjacent to the book, at a distance which doesn't make the text too small to read.

- In case of emergency, press the two halves of the book together <u>firmly</u> in order to close.

- Before attempting to use this book, familiarise yourself with the following <u>safety information</u>:

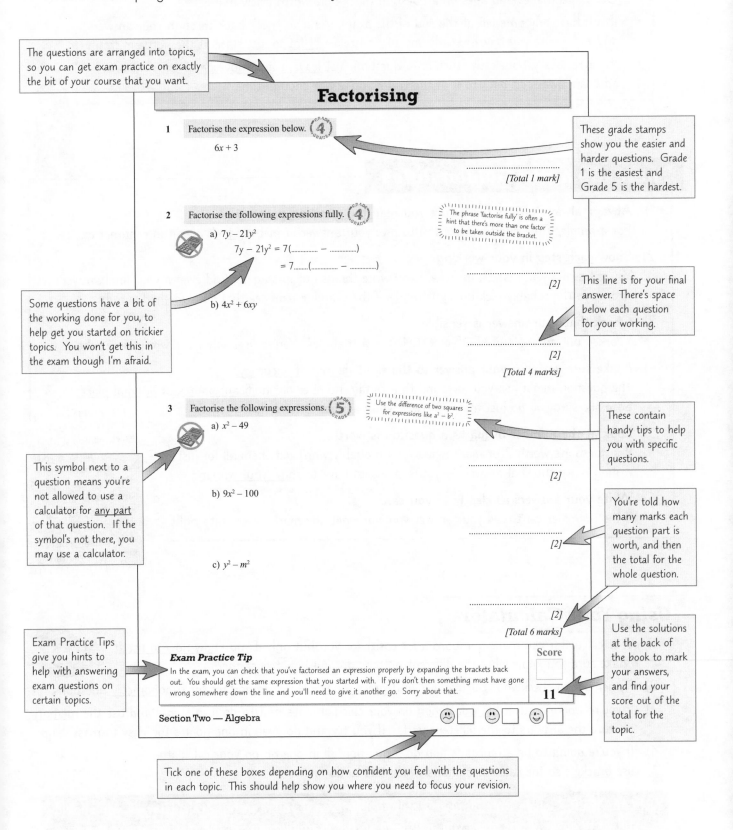

The questions are arranged into topics, so you can get exam practice on exactly the bit of your course that you want.

Factorising

1 Factorise the expression below. ④

$6x + 3$

....

[Total 1 mark]

These grade stamps show you the easier and harder questions. Grade 1 is the easiest and Grade 5 is the hardest.

The phrase 'factorise fully' is often a hint that there's more than one factor to be taken outside the bracket.

2 Factorise the following expressions fully. ④

a) $7y - 21y^2$

$7y - 21y^2 = 7(\text{.............} - \text{.............})$

$= 7\text{.......}(\text{.............} - \text{.............})$

b) $4x^2 + 6xy$

....

[2]

....

[2]

[Total 4 marks]

Some questions have a bit of the working done for you, to help get you started on trickier topics. You won't get this in the exam though I'm afraid.

This line is for your final answer. There's space below each question for your working.

3 Factorise the following expressions. ⑤

a) $x^2 - 49$

Use the difference of two squares for expressions like $a^2 - b^2$.

....

[2]

b) $9x^2 - 100$

....

[2]

c) $y^2 - m^2$

....

[2]

[Total 6 marks]

These contain handy tips to help you with specific questions.

This symbol next to a question means you're not allowed to use a calculator for <u>any part</u> of that question. If the symbol's not there, you may use a calculator.

You're told how many marks each question part is worth, and then the total for the whole question.

Exam Practice Tip

In the exam, you can check that you've factorised an expression properly by expanding the brackets back out. You should get the same expression that you started with. If you don't then something must have gone wrong somewhere down the line and you'll need to give it another go. Sorry about that.

Score

11

Section Two — Algebra

☹ ☐ ☺ ☐ ☺ ☐

Exam Practice Tips give you hints to help with answering exam questions on certain topics.

Use the solutions at the back of the book to mark your answers, and find your score out of the total for the topic.

Tick one of these boxes depending on how confident you feel with the questions in each topic. This should help show you where you need to focus your revision.

Exam Tips

Exam Stuff

Timings in the exam are really important, so here's a quick guide...

- Aim to spend about a <u>minute per mark</u> working on each question (i.e. 2 marks = 2 mins). Don't spend ages and ages on a question that's only worth a few marks.
- If you have any time left at the end of the exam, use it to <u>check</u> back through your answers and make sure you haven't made any silly mistakes. <u>Not</u> to just stare at the hottie in front.
- If you're totally, hopelessly stuck on a question, just <u>leave it</u> and <u>move on</u> to the next one. You can always <u>go back</u> to it at the end if you've got enough time.

There are a Few Golden Rules

1) **Always, always, always make sure you <u>read the question properly</u>.**
   For example, if the question asks you to give your answer in metres, <u>don't</u> give it in centimetres.

2) **Show <u>each step</u> in your <u>working</u>.**
   You're less likely to make a mistake if you write things out in stages. And even if your final answer's wrong, you'll probably pick up <u>some marks</u> if the examiner can see that your <u>method</u> is right.

3) **Check that your answer is <u>sensible</u>.**
   Worked out an angle of 450° or 0.045° in a triangle? You've probably gone wrong somewhere...

4) **Make sure you give your answer to the right <u>degree of accuracy</u>.**
   The question might ask you to round to a certain number of <u>significant figures</u> or <u>decimal places</u>. So make sure you do just that, otherwise you'll almost certainly lose marks.

5) **Look at the number of <u>marks</u> a question is worth.**
   If a question's worth 2 or more marks, you probably won't get them all for just writing down the final answer — you're going to have to <u>show your working</u>.

6) **Write your answers as <u>clearly</u> as you can.**
   If the examiner can't read your answer you won't get any marks, even if it's right.

Obeying these Golden Rules will help you get as many marks as you can in the exam — but they're no use if you haven't learnt the stuff in the first place. So make sure you revise well and do <u>as many</u> practice questions as you can.

Using Your Calculator

1) Your calculator can make questions a lot easier for you but only if you <u>know how to use it</u>. Make sure you know what the different buttons do and how to use them.

2) Remember to check your calculator is in <u>degrees mode</u>. This is important for <u>trigonometry</u> questions.

3) If you're working out a <u>big calculation</u> on your calculator, it's best to do it in <u>stages</u> and use the <u>memory</u> to store the answers to the different parts. If you try and do it all in one go, it's too easy to mess it up.

4) If you're going to be a renegade and do a question all in one go on your calculator, use <u>brackets</u> so the calculator knows which bits to do first.

REMEMBER: <u>Golden Rule number 2</u> still applies, even if you're using a calculator — you should still write down <u>all</u> the steps you are doing so the examiner can see the method you're using.

Types of Number and BODMAS

1 Work out: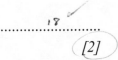

a) $11 + 14 \div 2$

$11 + 7 = 18$

............18 ✓

[2]

b) $(20 - 15) \times (4 + 6)$

$5 \times 10 = 50$ ✓

............50

[2]

[Total 4 marks]

2 Write down the value of each of the following:

a) 9^2 ①

............81 ✓

[1]

b) 4^3 ②

$4 \times 4 = 16$

$16 \times 4 = 64$ $\begin{array}{r} 216 \\ \times\ 4 \\ \hline 64 \end{array}$

............64 ✓

[1]

[Total 2 marks]

3 Circle the integers from the list below. ②

2.5   ⟨18⟩ ✓   ⟨$\frac{12}{6}$⟩ $= 2$ 1.333...   1.75   ⟨-22⟩ ✓ $\frac{7}{9}$

[Total 2 marks]
1

4 Use your calculator to work out $\dfrac{197.8}{\sqrt{0.01 + 0.23}}$. ②

Write down all the figures on your calculator display.

This question is worth two marks so you'll need to show some working as well as the final answer.

$\dfrac{197.8}{\sqrt{0.01 + 0.23}} = \dfrac{197.8}{\sqrt{0.24}}$

$= 403.7575593$ ✓

............403.7575593 ✓

[Total 2 marks]
1

Score: ☐ 8

10

Wordy Real-Life Problems

1 Jamie has 522 stickers. He gives 197 to his brother and 24 to his sister.
How many stickers does he have left?

```
 ¹¹97            522²
+  24          - 221
————           ————
 221             301
```  ✓

..........301.......✓

[Total 2 marks]

2 Eric goes to town with £15. He spends £8.50 on a new scarf.
He meets his nan who gives him £20 and tells him to take £10 of it home for his sister.
Eric then sees a jumper he likes which costs £18.

If Eric buys the jumper, will he still have £10 to give to his sister?
Show how you worked out your answer.

```
 ⁰¹⁴
 ₤5.'00         6.50          6.50
-   8.50       +20.00        +10.00
————————       ————————      ————————
   6.50         26.50         16.50
```

15 - 8.50 + 20 - 18 = ₤8.50 with sister's money   without

He will not have enough to give his sister if he buys
the jumper ✓

[Total 2 marks]

3 Parvati and Zayn stop at a café for breakfast.

New Road Café			
Breakfast Snacks:		*Drinks:*	
Toast (2 slices)	£1.50	Tea	£1.40
Yoghurt with fruit	£1.90	Coffee	£1.50
Raisin bagel	£2.30	Fresh orange juice	£1.80
Breakfast deal of the day:			
Raisin bagel & fresh orange juice	£3.40		

Parvati buys a raisin bagel and a cup of tea.
Zayn buys a yoghurt with fruit and a fresh orange juice. They each pay separately.

Work out how much money they would have saved if they had paid for everything together.

P = 2.30 + 1.40 = ₤3.70 ✓
Z = 1.90 + 1.80 = 3.70 ✓   } + = ₤7.40

7.40 - 6.70 = ₤0.70

3.40 + 1.40 + 1.90 = ₤6.70

They'd save 70p if they paid together

```
 ¹1.90          ¹3.40
+ 1.80         + 1.40
——————           1.90
  3.70         ——————
                ₤6.70
```

£ ..6.70.. ..0.70..

[Total 3 marks]

4 Sue and Alan meet Mark in a juice bar. Mark offers to buy a round of drinks.

 Mark wants a Passion Fruit Punch and Sue and Alan both want a Tutti Frutti.

Mark pays with a £10 note.
How much change will he get?

Juice Bar Price List	
St Clements:	£2.80
Cranberry Crush:	£2.90
Tutti Frutti:	£2.40
Passion Fruit Punch:	£2.15

2.15 + 2.40 + 2.40

2.15
+ 2.40
2.40
‾‾‾‾‾
6.95

⁹0.⁹⁸⁰⁰
- 6.95
‾‾‾‾‾
3.05

£ ......3.05......

[Total 2 marks]

5 Theo has a 500 ml bottle of a fizzy drink. Poppy has 216 ml of the same fizzy drink in a glass. Theo gives Poppy some of his drink so that they each have the same amount.

How much drink does Theo give to Poppy?

500 + 216 = 716 ml

716 ÷ 2 = 358 ml each

500 - 358 = 142 ml

......142...... ml

[Total 2 marks]

6 Georgie is a sales representative. She drives to different companies to sell air conditioning units.

When she has to travel, her employer pays fuel expenses of 30p per mile. She drives to a job in the morning and drives home again later that day. She is also given £8 to cover any food expenses for each day that she is not in the office.

Monday: Buckshaw, 30 miles
Tuesday: in office
Wednesday: Wortham, 28 miles
Thursday: Harborough, 39 miles
Friday: Scotby, 40 miles

The distances to her jobs for this week are shown on the right.

Find Georgie's total expenses for this week.

Hint: think carefully about the total distance travelled.

M: 30
×0.30
0.30
× 30
‾‾‾‾
000
0900
‾‾‾‾
9.00

£9 for fuel
+ 8
= £17 for Monday

20 + 24.70 + 16.40
+ 17 = 78.10

W: ×0.30
× 28
240
0600
‾‾‾‾
08.40

8.40
+ 8
‾‾‾‾
£16.40

£8.40 for fuel

Thurs: ×0.30
× 39
770
0900
‾‾‾‾
£16.70 for fuel
+ 8
‾‾‾‾
£24.70

F: 0.30
× 40
000
1200
‾‾‾‾
£12.00
+ 8
‾‾‾‾
£20

£ ......78.10......

[Total 4 marks]

Score: 5

15

Multiplying and Dividing

1   Write a number in each box below to make the calculation correct.

$$12 \times 15 = 3 \times \boxed{60} = \boxed{180}$$

(handwritten:
12
× 15
60
1 2 0
1 8 0 *)*

[Total 2 marks]

2   Milk chocolate monsters cost 38p each and white chocolate witches cost 44p each.
A shop sold 468 milk chocolate monsters and 402 white chocolate witches.

How much more was spent on milk chocolate monsters than white chocolate witches in total?

(handwritten working:
468   0.38   MC: 468 × 0.38 = £177.84
× 468
WC: 402 × 0.44 = 176.88
177.84 − 176.88 = 0.96 *)*

£ .....0.96.....

[Total 3 marks]

3   Four numbers multiply together to give 672. Three of the numbers are 2, 6 and 7.

What is the fourth number?

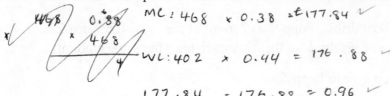

(handwritten working:
2 × 6 × 7
12 × 7 = 84
8
84 | 672
3 84
× 8
672
2 84   2 84
× 5   × 7
420   588 *)*

.....8.....

[Total 3 marks]

4   Alanna buys 15 tickets for a concert for herself and some friends.
Each ticket is the same price. She pays with £200 and gets £5 change.

How much does each ticket cost?

(handwritten working:
200 − 5 = £195 in total
33
15 | 195
13
15 | 195
15
45 *)*

£ .....33.13.....

[Total 3 marks]

5   A school is organising a trip for 29 Year 10 students and 57 Year 11 students.

a) Bus tickets cost £1.90 each. How much will bus tickets
for all the Year 10 students cost?

(handwritten working:
1.90 × 29
£1.90
× 29
1710
3800 = 55.10 *)*

£ .....55.10.....

[2]

b) The students will be divided up into groups of no more than 6.
What is the minimum possible number of groups?

(handwritten working:
29 + 57 = 86
86 ÷ 6 =
14.33 *)*

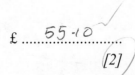

.....15.....

[2]

[Total 4 marks]

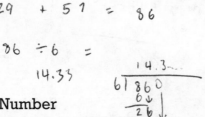

6 In an office, each worker eats 2 biscuits per day. There are 20 workers in the office every day of the week. Biscuits come in packets of 14 and cost £1.30 per packet. ②GRADE

a) How many biscuits are eaten in the office per week?

20 × 2 = 40 biscuits per day

40 × 7 = 280 biscuits/week

.... 280 40 biscuits

[2]

b) How much does the office spend on biscuits per week?

40 ÷ 14 = 2.857142

= 3

14⟌140

280 ÷ 14 = 20 packets/week

1·30
× 3
———
3.90

20 × 1·30 = 26/week

£ 26 3.90

[2]

[Total 4 marks]

7 Work out how many of each individual item below were sold if: ③GRADE

a) Sophie spent £14 on rulers that cost £0.70 each.

20
0·70⟌1400
140
0 0
0

.... 20 ✓

[2]

b) Michael spent £2.76 on pencils that cost £0.12 each.

23
0·12⟌2·76
24
36
36
0

.... 23 ✓

[2]

[Total 4 marks]

8 James is having a party. He has worked out how much food he needs to buy per person and goes to the shop to buy the things that he needs. ③GRADE

Crisps come in 300 g packets.

There are 8 slices per pizza.

James is expecting there to be 15 people, including himself, at the party.

For each person:
• 3 slices of pizza
• 25 g of crisps

How many pizzas and how many packets of crisps does he need to buy?

15
× 3
———
45

45 slices of pizza in Total ✓

25
× 15
———
125
250
———
375

375 g of crisps in total ✓

.... 6 pizzas

.... 2 packets of crisps

[Total 5 marks]

5.625
8⟌45000
40
50
48
20
16
40

= 6 pizzas

1·25
300⟌37500
300
750
600
1500
1500
2

= 2 packets

Score: 19

28

Section One — Number

Negative Numbers

1 Put the numbers below in order from lowest to highest.

 ~~-8~~ ~~-3~~ ~~-3~~ ~~7~~ ~~0~~ ~~-9~~ 10

-9 , -8 , -3 , 0 , 3 , 7 , 10

[Total 1 mark]

2 Three numbers multiply to give 288. Two of the numbers are –3 and 12.

What is the third number?

-3 × 12 = -36

-36 ⟌ 288 = -8

²36 × 4 = 144
3 36 × 6 = 216
4 36 × 7 = 252
4 36 × 8 = 288

-8

[Total 3 marks]

3 Put the numbers below in order from lowest to highest.

 ~~-0.75~~ ~~-0.23~~ ~~-0.61~~ ~~-0.35~~ 1.06 ~~-1.12~~

-1.12 , -0.61 , -0.23 , 0.35 , 0.75 , 1.06

[Total 1 mark]

4 Esther has three number cards.

| -4 | 3 | 5 |

She is going to use the numbers to make a new number.
She can use the operations +, –, ×, ÷ and brackets.
What is the highest number she can make?

-4 × 3 × 5 = -60

-4 × 3 + 5 = 4

-4 -

(3 - -4) × 5 = 7 × 5 = 35

35

[Total 3 marks]

Score: 5 / 8

Prime Numbers

1 Circle all the prime numbers in the list below. (3)

27   (29)   (31)   33   35   (37)   39

[Total 2 marks]

2 Write down **one** prime number that lies between each of the following pairs of numbers: (3)

a) 45 and 55

...........51...........
47/53   *[1]*

b) 62 and 72

...........67...........
[1]

$3\overline{)67}$

[Total 2 marks]

3 Look at the list of numbers below. (3)

1   7   11   12   15   21

a) Write down a number from the list which is a prime number.

...........7...........
[1]

b) Write down two numbers from the list whose sum is a prime number.

...........7........... and ....12...........
[1]

[Total 2 marks]

4 Jack says, "there are no prime numbers between 100 and 110." (3)
Is he correct? Give evidence for your answer.

Jack is incorrect

Prime numbers = (101, 103, 107, 109)

[Total 2 marks]

5 Jay thinks of a prime number. The sum of its digits is one more than a square number. (3)

Write down one number Jay could be thinking of.

12
1 + 2 = 3 - 2 = 1

37 (3 + 7 = 10, which is one more than 9, a square number)

...........12...........
[Total 2 marks]

Score: 5
/ 10

Section One — Number

Multiples, Factors and Prime Factors

1 Look at the list of numbers below. **(3)**

80   66   64   72   62   74

a) Write down a number from the list that is a multiple of 12.

........................72........................
[1]

b) Write down a number from the list that is a factor of 128.

........................64........................
[1]

c) Write down a number from the list that is a multiple of 5 and a multiple of 4.

........................80........................
[1]

[Total 3 marks]

2 Write down: **(3)**

a) **all** the factors of 28,

1 × 28 , 2 × 14 , 4 × 7 ,

........1, 2, 4, 7, 14, 28........
[2]
|

b) all the multiples of 8 which appear in the list below.

~~55~~ 56 ~~57~~ ~~58~~ 59 60 ~~61~~ 62 ~~63~~ ~~64~~ 65

........56, 64 ✓........
[1]

[Total 3 marks]

3 Write 72 as a product of its prime factors. **(4)**

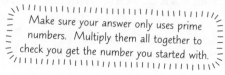
Make sure your answer only uses prime numbers. Multiply them all together to check you get the number you started with.

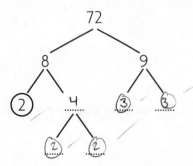

$72 = 2 \times 2 \times 2 \times 3 \times 3$

........2 × 2 × 2 × 3 × 3........

[Total 2 marks]

Score: **7**

8

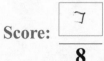

LCM and HCF

1 Find:

a) the lowest common multiple of 15 and 20,

15, 30, 45, <u>60</u>

20, 40, 60

.... 60 ✓

[2 mark]

b) the highest common factor of 42 and 70.

42

1 × 42, 2 × 21, 3 × 14, 6 × 7,   1 × 70, 2 × 35, 5 × 14

7 × 10,

.... 14 ✓

[2 mark]

[Total 4 marks]

2 $P = 3^7 \times 11^2$ and $Q = 3^4 \times 7^3 \times 11$.

Write as the product of prime factors:

a) the LCM of P and Q,

P = 3 × 3 × 3 × 3 × 3 × 3 × 3

9 × 9 × 9 × 3

9 × 9

81 × 27

81 × 27 = 2187 × 121

b) the HCF of P and Q.

[1 mark]

[1 mark]

[Total 2 marks]

3 Phil is making jam.

He needs to buy mini jam jars which come in <u>packs of 12</u>, lids which come in <u>packs of 16</u> and labels which come in <u>packs of 36</u>. He doesn't want to have any items left over.

Find the smallest number of packs of each item he can buy.

Multiples of 12 are: 12, 24, 36, 48, 60, 72, 84, 96, 108, 120, 132, 144, 156.
Multiples of 16 are: 16, 32, 48, 64, 80, 96, 112, 128, 144, 160
Multiples of 36 are: 36, 72, 108, 144, 180, ..., ..., ...

The LCM of 12, 16 and 36 is 144, which is the minimum number of each item he needs.
The minimum number of packs of jars he needs is 144 ÷ 12 = 12 packs
The minimum number of packs of lids he needs is 144 ÷ 16 = 9 packs
The minimum number of packs of labels he needs is 144 ÷ 36 = 4 packs

........12........ packs of jars, ...9........ packs of lids and ....4........ packs of labels

[Total 3 marks]

Score: 4

9

Section One — Number

Fractions

1 Find:

 a) $\frac{3}{5}$ of 60,

$\frac{3}{5} \times 60 = \frac{180}{5} = 36$

$5\overline{)18^30}$ $\frac{36}{}$

..............36.............

[2]

b) 15 out of 40 as a fraction in its simplest form.

$^{15}/_{40} = {}^3/_8$

.............3/8.............

[2]

[Total 4 marks]

2 Work out the following: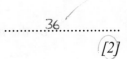

 a) $\frac{1}{2} \times \frac{1}{6}$ $\frac{1}{2} \times \frac{1}{6} = \frac{1 \times \dots}{12 \times \dots} = \frac{\dots}{\dots}$

.............¹/₁₂.............

[1]

b) $\frac{2}{3} \div \frac{3}{5}$ $\frac{2}{3} \div \frac{3}{5} = \frac{2}{3} \times \frac{5}{3} = \frac{\dots \times \dots}{\dots \times \dots} = \frac{10}{9}$

$= 1\frac{1}{9}$

.............1 ¹/₉.............

[2]

[Total 3 marks]

3 Which of these fractions is closest to 1?

$\frac{5}{6}$ $\frac{3}{4}$ $\frac{7}{8}$

Next time, make the denominators the same,
then it'll become easier

.............7/8.............

[Total 1 mark]

4 Half of the rectangle shown is shaded. The other half is split into 9 equal squares.
Aito says, "If I shade two more squares, $\frac{1}{2} + \frac{2}{18} = \frac{3}{20}$ of the rectangle will be shaded."
How can you tell, without doing the calculation, that Aito is wrong?

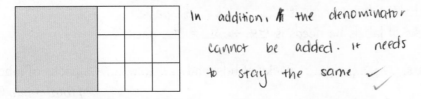

In addition, the denominator
cannot be added. It needs
to stay the same ✓

[Total 1 mark]

Section One — Number

5 The number of people at last Saturday's Norchester City game was 12 400. Season ticket holders made up $\frac{3}{8}$ of the crowd. How many season ticket holders were there?

$$\frac{3}{8} \times \frac{12\,400}{1} = \frac{37200}{8}$$

$$= 4650$$

4650 ✓ ticket holders

[Total 2 marks]

---

6 Francis owns all the shares of his company.
He sells $\frac{2}{15}$ of the shares to Spencer and $\frac{5}{12}$ of the shares to Jamie.

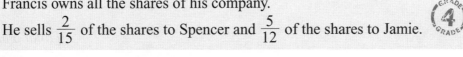

What fraction of the shares does Francis still own? Give your answer in its simplest form.

$$\begin{array}{r} 12 \\ \times\ 15 \\ \hline 60 \\ 120 \\ \hline 180 \end{array}$$

$$\frac{24}{180} + \frac{75}{180} = \frac{99}{180}$$

$$\begin{array}{r} 15 \\ \times\ 5 \\ \hline 15 \end{array}$$

$$\begin{array}{r} 180 \\ -\ 99 \\ \hline 81 \end{array}$$ Francis still owns $\frac{81}{180}$ of the shares

$$= \frac{9}{20}$$

9/20 ✓

[Total 3 marks]

---

7 *ABC* is an equilateral triangle. It has been divided into smaller equilateral triangles as shown below.

What fraction of triangle *ABC* is shaded?

Find the shaded amounts separately then add them up

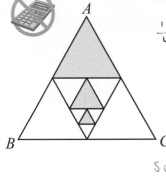

$$\frac{1}{4} + \frac{1}{4} + \frac{1}{3}$$

$$\frac{3}{12} + \frac{3}{12} + \frac{4}{12} = \frac{10}{12}$$

Shaded regions are $\frac{1}{4}$, $\frac{1}{4} \times \frac{1}{4} = \frac{1}{16}$ and $\frac{1}{4} \times \frac{1}{4} \times \frac{1}{4} = \frac{1}{64}$

So total area shaded $= \frac{1}{4} + \frac{1}{16} + \frac{1}{64} = \frac{16}{64}$

$$+ \frac{4}{64} + \frac{1}{64} = \frac{21}{64}$$

$^{10}/_{12}$ $^{21}/_{64}$

[Total 3 marks]

---

8 A factory buys 25 tonnes of flour. $17\frac{1}{2}$ tonnes of the flour is used to make scones.
$\frac{1}{5}$ of the scones are cheese scones.

What fraction of the total amount of flour is used to make cheese scones?

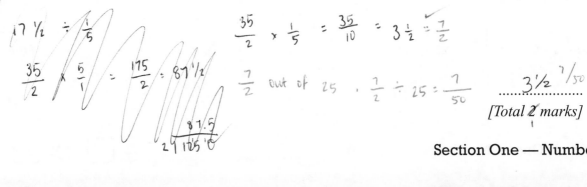

$$17\frac{1}{2} \div \frac{1}{5}$$

$$\frac{35}{2} \times \frac{5}{1} = \frac{175}{2} = 87\frac{1}{2}$$

$$\frac{35}{2} \times \frac{1}{5} = \frac{35}{10} = 3\frac{1}{2} = \frac{7}{2}$$

$\frac{7}{2}$ out of 25 . $\frac{7}{2} \div 25 = \frac{7}{50}$

$3\frac{1}{2}$ $^{7}/_{50}$

[Total 2 marks]

Section One — Number

9   Work out the following, giving your answers as mixed numbers in their simplest form:

a) $1\frac{1}{8} \times 2\frac{2}{5}$

$$\frac{9}{8} \times \frac{12}{5} = \frac{108}{40} = 2\frac{28}{40} = 2\frac{14}{20} = 2\frac{7}{10}$$

$\frac{960}{80}$ ... $\frac{-80}{28}$

.......$2\frac{7}{10}$.......

[3]

b) $1\frac{3}{4} \div \frac{7}{9}$

$$\frac{7}{4} \div \frac{7}{9} = \frac{7}{4} \times \frac{9}{7} = \frac{63}{28} = 2\frac{7}{28} = 2\frac{1}{4}$$

$\frac{28}{28} / \frac{56}{56}$ $\frac{56{\cdot}3}{-56} / \frac{}{7}$

.......$2\frac{1}{4}$.......

[3]

[Total 6 marks]

10   Scott and his 4 friends eat $\frac{5}{6}$ of a pizza each. Pizzas cost £4.50 each, or 2 for £7.

What is the minimum amount they will have to spend on pizzas?

$$\frac{5}{6} \times 4 = \frac{20}{6} = 3\frac{2}{6} = 3\frac{1}{3} \approx \text{around } 4 \text{ pizzas in total}$$

$\begin{array}{r} 4.50 \\ \times \quad 4 \\ \hline £18.00 \end{array}$ or $2 \times 7 = £14$

$5 \times \frac{5}{6} = \frac{25}{6} = 4\frac{1}{6} \rightarrow 5 \text{ pizzas in total}$

£ ......14......

[Total 3 marks]

$\text{cost} = (2 \times 7) + 4.50 = 14 + 4.50 = £18.50$

11   Mr Fletcher owns 36 acres of land. He uses $\frac{5}{12}$ of his land for wheat, $\frac{1}{3}$ for cows and $\frac{1}{6}$ for pigs. The remaining $\frac{1}{12}$ of the land is taken up by a farmhouse and garden.

Each acre of land used for wheat, cows or pigs costs £400 per year.
An acre used for wheat makes £1100 per year.
An acre used for cows makes £1450 per year.
An acre used for pigs makes £1250 per year.

Work out Mr Fletcher's total profit per year.

$$\frac{5}{12} \times 36 = \frac{180}{12} = 15 \text{ acres for wheat} \checkmark$$

$\text{cows} = \frac{1}{3} \times 36 = \frac{36}{3} = 12 \text{ acres} \checkmark$

$\text{pigs} = \frac{1}{6} \times 36 = \frac{36}{1} = \frac{36}{6} = 6 \text{ acres} \checkmark$

$\text{cost to run/year} = (15 + 12 + 6) \times 400 \checkmark$

$\text{garden/farmhouse} = \frac{1}{12} \times 36 = \frac{36}{12} = 3 \text{ acres}$

£ ......21600 28200......

[Total 4 marks]

$(400 \times \cancel{36} \; 33) + (1100 \times 15) + (1450 \times 12) + (1250 \times 6)$

$16500 \quad + \quad 17400 \quad + \quad 7500$

£13 200 $= £21600$

$- 13200$

$= 28200$

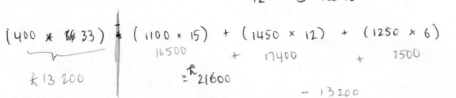

Score: $\boxed{21}$

32

Section One — Number

Fractions, Decimals and Percentages

1 Convert each of the following: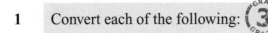

a) $\frac{3}{4}$ to a decimal,

$3\overline{)3.^30^20}$

.... 0.75 ✓

[1]

b) 0.06 to a percentage.

$^6/_{100}$

.........6......%

[1]

[Total 2 marks]

2 Circle the value of the digit 6 in the number 4.961.

$\frac{6}{10}$ 6 $\boxed{\frac{6}{100}}$ $\frac{1}{60}$ $\frac{6}{1000}$

[Total 1 mark]

3 Write the following in order of size, starting with the smallest.

Start by writing all the numbers as decimals.

65%² $\frac{2}{3}$ ⁴   0.065 ¹ $\frac{33}{50}$ ³

0.65   0.666   0.66

$50\overline{)\begin{array}{c}0.66\\ 330^30\\ 300\downarrow\\ \hline 300\end{array}}$

$3\overline{)\begin{array}{c}0.66\\ 200\end{array}}$

.....0.065...... , ...65%...... , ...$^{33}/_{50}$...... , ...$^2/_3$......

[Total 2 marks]

4 Samuel, Eli, Robert and Jenny split the bill at a restaurant. Samuel pays $\frac{1}{4}$ of the bill and Eli and Robert each pay 20% of the bill. Jenny pays £17.50.

 How much was the bill in total?

Samuel = $\frac{1}{4}$ Eli & Robert = $\frac{2}{10}$

$\frac{1}{4} + \frac{2}{10} + \frac{2}{10} = \frac{10}{40} + \frac{8}{40} + \frac{8}{40} = \frac{26}{40}$

$17.50 = 17\frac{1}{2}$

$\frac{26}{40} + \frac{35^{\times 20}}{2^{\times 20}} = \frac{26}{40} + \frac{700}{40} = \frac{726}{40}$

£ ...... 18.15 ......

[Total 4 marks]

£18.15

$\begin{array}{c}35\\ \times 20\\ \hline 00\\ 700\\ \hline 700\end{array}$

$40\overline{)\begin{array}{c}18.15\\ 72600\\ 40\downarrow\\ \hline 326\\ 320\downarrow\\ \hline 60\\ 40\downarrow\\ \hline 200\end{array}}$

Score: ☐

9

Section One — Number

 ☐   ☐   ☺ ☐

Rounding

1 Round the following to the given degree of accuracy.

   a) Josh has 123 people coming to his party.
      Write this number to the nearest 10.

      ....120............
      [1]

   b) The attendance at a football match was 2568 people.
      What is this to the nearest hundred?

      ....2600............
      [1]

   c) The population of Ulverpool is 452 529.
      Round this to the nearest 100 000.

      ....500 000............
      [1]

      [Total 3 marks]

2 The distance between two stars is 428.6237 light years.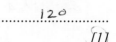

   a) Round this distance to one decimal place.

      ....428.6............ light years
      [1]

   b) Round this distance to 2 significant figures.

      ....428.62............ light years
      430
      [1]

      [Total 2 marks]

3 Use your calculator to find:

$$\frac{4.32^2 - \sqrt{13.4}}{16.3 + 2.19}$$

   Give your answer to 3 significant figures.

   $$\frac{4.32^2 - \sqrt{13.4}}{16.3 + 2.19} = \frac{4.32^2 - \sqrt{13.4}}{18.49}$$

   $$= 0.8113466174$$

   ....0.811............
   [Total 2 marks]

4 A number rounded to the nearest whole number is 122.

   What is the smallest possible value of the number?

   ....121.5............
   [Total 1 mark]

   Score: [5]

   8

Estimating and Error

1 The man in this picture is 180 cm tall.

Use this information to estimate the height of the penguin.

$$\approx 180 \div 3 = 60 \text{ cm}$$

.......120..... cm

[Total 2 marks]

2 A stall sells paperback and hardback books. Paperback books cost £4.95 and hardback books cost £11. One Saturday, the stall sells 28 paperback and 19 hardback books.

a) Find an estimate for the amount of money the stall made that day.
   Show all your working.

```
      4.95              11            138.60
   ×    28           ×  19         +  209
   ─────────         ─────────        ──────────
    139  80            99            347.60
    9900             110
   ─────────         ─────────
   £ 138.60          £ 209
```

£ ....347.60.......

[2]

b) The actual amount the stall made was £347.60.
   Do you think your estimate was sensible? Explain your answer.

....Since I did all of the working, I got the same exact answer....

[1]

[Total 3 marks]

3 There are 1750 green pens in a warehouse. Each pen weighs 12 g.

a) Find an estimate for the weight of all the green pens in the warehouse. Give your answer in kg.

```
      1750
   ×    12
   ──────────
    3500
   17500
   ──────────
   21000
```

.......21000..... kg

[2]

b) Could an average person lift all the green pens in the warehouse? Explain your answer.

....No because 21000 kg is not possible for someone to carry something there heavy.....

[1]

[Total 3 marks]

Section One — Number

4   Estimate the value of $\dfrac{12.2 \times 1.86}{0.19}$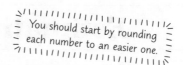

$$\dfrac{12 \times 2}{\underset{0}{0.19}} = \dfrac{24}{\underset{0}{0.19}} = 24$$

.... 24 ....

[Total 2 marks]

5   Joseph is weighing himself. His scales give his weight to the nearest kilogram.

According to his scales, Joseph is 57 kg.
What are the minimum and maximum weights that he could be?

Minimum weight: ......56.5...... kg

Maximum weight: ......57.45...... kg

[Total 2 marks]

6   The numbers *a* and *b* below have been rounded to 1 decimal place. Write down the error interval for each of the numbers.

a) *a* = 3.8

     Smallest possible value of a = 3.8 − ...0.05... = ...3.75...

     Largest possible value of a = 3.8 + ...0.05... = ...3.85...

     So error interval is: ...3.75... ≤ a < ...3.85...

....

[2]

b) *b* = 100.0

smallest : b = 100.0 − 0.05 = 99.95

largest : b = 100.0 + 0.05 = 100.05

error interval is

99.95 ≤ b < 100.05

....

[2]

[Total 4 marks]

Exam Practice Tip

Finding minimum and maximum values can be a bit tricky, as the amount you have to add and subtract changes depending on what the number's been rounded to. Remember, you always add and subtract <u>half</u> of the rounding unit (so if you were rounding to the nearest 10, the rounding unit is 10 ÷ 2 = 5).

Score

16

Powers and Roots

1 Use your calculator to find the following:

a) 8.7^3

658.503

[1]

b) $\sqrt[4]{1296}$ **2**

6

[1]

c) 4^{-2} **4**

0.0625   *¹/₁₆*

[1]

[Total 3 marks]

2 A square has an area of 6.25 cm². Find the length of one side of the square. **3**

$6.25 \div 2 = 3.125$

$\sqrt{6.25} = 2.5$ cm

.... cm
[Total 1 mark]

3 Simplify the expression $\dfrac{3^4 \times 3^7}{3^6}$. Leave your answer in index form. **5**

$\dfrac{3^{11}}{3^6} = 3^{(11-6)} = 3^5$

3^5
[Total 2 marks]

4 Work out the value of:

a) $6^5 \div 6^3$ **4**

$6^2 = 36$

36

[1]

b) $(2^4 \times 2^7) \div (2^3 \times 2^2)^2$ **5**

$(2^4 \times 2^7) = 2^{(4 + 7)} = 2^{11}$

$(2^3 \times 2^2) = 2^{(3 + 2)} = 2^5$, so $(2^3 \times 2^2)^2 = (2^5)^2 = 2^{10}$

So $(2^4 \times 2^7) \div (2^3 \times 2^2)^2 = 2^{11} \div 2^5 = 2^6 = $ ....

64

[2]

[Total 3 marks]

$2 \quad 2 \quad\quad 2 \quad 2 \quad 2 \quad 2$

$4 \quad\times\quad 4 \quad\times\quad 4$

$16 \quad\times\quad 4 \quad = 64$

$\begin{array}{r} 16 \\ \times\ 4 \\ \hline 64 \end{array}$

Score: ☐

9

 ☐   ☐   ☐

Section One — Number

Standard Form

1 $A = 4.834 \times 10^9, B = 2.4 \times 10^5, C = 5.21 \times 10^3$

 a) Write A as an ordinary number.

.....4834000000.....
[1]

b) Put A, B and C in order from smallest to largest.

A : 4834000000

B : 240000

C : 5210

.....C....., ...B....., ...A.....
[1]

[Total 2 marks]

2   The table on the right shows the masses of four different particles.

 a) Which particle is the heaviest?

.....C.....
[1]

b) What is the mass of particle C?
Give your answer as an ordinary number.

.....0.000014..... g
0.0000014
[1]

Particle	Mass (g)
Particle A	2.1×10^{-7}
Particle B	8.6×10^{-8}
Particle C	1.4×10^{-6}
Particle D	3.2×10^{-7}

c) How much more does particle D weigh than particle A?
Give your answer in standard form.

D : 0.00000032

A : 0.0000021

0.00000011

.....1.1×10^{-7}..... g
[2]

[Total 4 marks]

3   Light travels at approximately 2×10^5 miles per second.
The distance from the Earth to the Sun is approximately 9×10^7 miles.

How long will it take light to travel this distance?

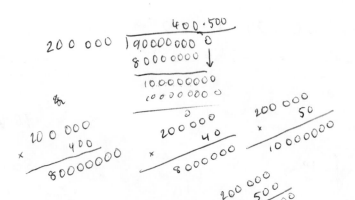

Light : 200000

E → S : 90000000

time = distance ÷ speed
= $(9 \times 10^7) \div (2 \times 10^5)$

450   400.5
400.5 × 10
.....seconds
[Total 2 marks]

Score: 5

8

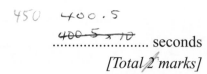

Algebra — Simplifying

1   Circle the simplified version of $4s - 3s + 9s$.

       $16s$ $12s$ ⟨$10s$⟩ $-8s$ $11s$

[Total 1 mark]

2   Select the correct words from the box to complete the sentence below.

root	identity	factor	equation
term	multiple		expression

     $3x$ is a ....*term*........... in the ....*expression*.......... $3x + 4y + 7$.

[Total 2 marks]

3   Simplify the following.

   a) $p + p + p + p$

                                          ....$4p$...........

                                              [1]

   b) $m + 3m - 2m$

                                          ....$2m$...........

                                              [1]

   c) $7r - 2p - 4r + 6p$

     $7r - 4r - 2p + 6p$

     $3r - 8p$ ....$3r - 8p$...........

                                      $4p + 3r$ *[2]*

[Total 4 marks]

4   Write the following in their simplest form.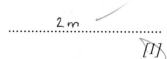

   a) $5pq + pq - 2pq$

                                          ....$4pq$...........

                                              [1]

   b) $2x^2 + 8x - 4x - x^2$

     $2x^2 - x^2 + 8x - 4x$

     $x^2 + 4x$ ....$x^2 + 4x$...........

                                              [2]

[Total 3 marks]

Score: 6

10

Algebra — Multiplying and Brackets

1 Simplify the following: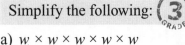

a) $w \times w \times w \times w \times w$

.... w^5 ✓ ....

[1]

b) $2a \times 5b$

.... $10ab$ ....

[1]

c) $8a^2 \div 4a$

.... $2a$ ....

[1]

[Total 3 marks]

2 Expand and simplify where possible.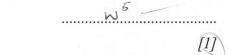

a) $3(x-1) + 5(x+2)$

$3x - 3 + 5x + 10$ ✓

$3x + 5x + 10 - 3 = 8x + 7$ ✓

.... $8x + 7$ ✓ ....

[2]

b) $4a(a + 2b)$

$4a^2 + 8ab$

.... $4a^2 + 8ab$ ✓ ....

[1]

c) $9 - 3(x + 2)$

$9 - 3x + 6$

$9 - 3x - 6 = 3 - 3x$

.... $15 - 3x$ ....

[2]

[Total 5 marks]

3 On the diagram below, shade the area represented by $pr + qs$.

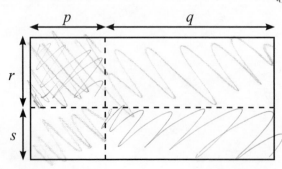

[Total 1 mark]

4 Show that $5(p + 6) - 2(p + 10) = 3p + 10$. **(3)**

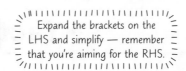

Expand the brackets on the LHS and simplify — remember that you're aiming for the RHS.

$5p + 30 - 2p + 20$

$5p - 2p + 30 + 20$

$3p + 50$

$5(p+6) - 2(p+10) = 5p + 30 - 2p - 20$

$= 3p + 10$

[Total 2 marks]

5 Expand and simplify the following:

a) $(x + 2)(x + 4)$ **(4)**

$(x \times x) + (x \times 4) + (2 \times x) + (2 \times 4)$

$x^2 + 4x + 2x + 8$ $x^2 + 4x + 2x + 8$

$x^2 + 6x + 8$

.... $x^2 + 6x + 8$ ✓

[2]

b) $(y + 3)(y - 3)$ **(4)**

$y^2/4$

$y^2 - 3y + 3y - 9$

$y^2 - 9$

.... $y^2 - 9$ ✓

[2]

c) $(2z - 1)(z - 5)$ **(5)**

$2z^2 - 10z - 1z + 5$ ✓

$2z^2 - 9z + 5$

$2z^2 - 11z + 5$

.... $2z^2 - 9z + 5$

[2]

[Total 6 marks]

6 Expand and simplify the following:

a) $(a - 7)^2$ **(4)**

$(a-7)(a-7)$

$(a \times a) + (a \times -7) + (-7 \times a) + (-7 \times -7)$

$a^2 \quad -7a - 7a + 49 = a^2 - 14a + 49$

.... $a^2 - 14a + 49$

[2]

b) $(3b + 2)^2$ **(5)**

$(3b + 2)(3b + 2)$

$(3b \times 3b) + (3b \times 2) + (2 \times 3b) + (2 \times 2)$

$3b^2 + 6b + 6b + 4$

$3b^2 + 12b + 4$

.... $3b^2 + 12b + 4$

[2]

[Total 4 marks]

Exam Practice Tip

When you're multiplying out two brackets, just remember to multiply everything in the first bracket by everything in the second bracket. And if you have a squared bracket, always always always write it out as two brackets first — otherwise you're more likely to make a mistake and lose marks.

Score

21

Factorising

1   Factorise the expression below.

$6x + 3$

$3(2x + 1)$

$$\text{................} 3(2x + 1) \checkmark$$

[Total 1 mark]

2   Factorise the following expressions fully.

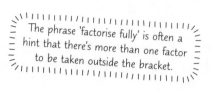
The phrase 'factorise fully' is often a hint that there's more than one factor to be taken outside the bracket.

 a) $7y - 21y^2$

$$7y - 21y^2 = 7(\underline{\quad y \quad} - \underline{\quad 3y^2 \quad})$$
$$= 7\underline{y}(\underline{\quad 1 \quad} - \underline{\quad 3y \quad})$$

$$\text{................} 7y(1 - 3y) \checkmark$$

[2]

b) $4x^2 + 6xy$

$$4x \cancel{/} 6x$$
$$2x(2x +$$
$$2(2x^2 + 3xy$$
$$= 2x(2x + 3y)$$

$$\text{................} 2x(2x + 3y)$$

[2]

[Total 4 marks]

3   Factorise the following expressions.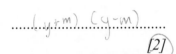

Use the difference of two squares for expressions like $a^2 - b^2$.

 a) $x^2 - 49$

$$x \cancel{/} x$$
$$x^2 - 7^2 = (x + 7)(x - 7)$$

$$\text{................}$$

[2]

b) $9x^2 - 100$

$$(3x)^2 - 10^2 = (3x + 10)(3x - 10)$$

$$\text{................}$$

[2]

c) $y^2 - m^2$

$$(y + m)(y - m) \checkmark$$

$$\text{................} (y + m)(y - m)$$

[2]

[Total 6 marks]

Exam Practice Tip

In the exam, you can check that you've factorised an expression properly by expanding the brackets back out. You should get the same expression that you started with. If you don't then something must have gone wrong somewhere down the line and you'll need to give it another go. Sorry about that.

Score

5

11

Solving Equations

1 Solve these equations for x.

a) $x + 3 = 12$

$\underline{-3 \quad -3}$

$x = 9$

$x = \dots\dots 9 \dots\dots$ ✓

[1]

b) $6x = 24$

$\dfrac{6x}{6} \quad \dfrac{24}{6}$

$x = 4$

$x = \dots\dots 4 \dots\dots$ ✓

[1]

c) $\dfrac{x}{5} = 4$

$5 \cdot \dfrac{x}{5} = 4 \times 5$

$x = 20$

$x = \dots\dots 20 \dots\dots$ ✓

[1]

[Total 3 marks]

2 Solve the equations below.

a) $p - 11 = -7$

$\underline{+11 \quad +11}$

$p = 4$

$p = \dots\dots 4 \dots\dots$ ✓

[1]

b) $2y - 5 = 9$

$\underline{+5 \quad +5}$

$\dfrac{2y}{2} = \dfrac{14}{2}$

$y = 7$

$y = \dots\dots 7 \dots\dots$ ✓

[2]

c) $3z + 2 = z + 15$

$3z - 2 = 15 - 2$

$\dfrac{2z}{2} = \dfrac{13}{2}$

$z = 6.5$

$\begin{array}{r} 6.5 \\ 2\overline{\smash{)}13.0} \\ \underline{12} \\ 10 \\ \underline{10} \\ 0 \end{array}$

$z = \dots\dots 6.5 \dots\dots$ ✓

[2]

[Total 5 marks]

3 Solve the following equations.

a) $40 - 3x = 17x$

$\underline{+3x \quad +3x}$

$\dfrac{40}{20} = \dfrac{20x}{20} \qquad x = 2$

$x = \dots\dots 2 \dots\dots$ ✓

[2]

b) $2y - 5 = 3y - 12$

$-5 + 12 = 3y - 2y$

$7 = y$

$y = \dots\dots 7 \dots\dots$ ✓

[2]

[Total 4 marks]

Section Two — Algebra

4 Find the solution to each of the following equations.

a) $3(a + 2) = 15$

$$3a + 6 = 15$$
$$ -6 \quad -6$$
$$\overline{}$$
$$\frac{3a}{3} = \frac{9}{3}$$
$$a = 3$$

$a = $ ....3................... ✓

[3]

b) $5(2b - 1) = 4(3b - 2)$

$$10b - 5 = 12b - 8$$

$$-5 + 8 = 12b - 10b$$

$$\frac{3}{2} = \frac{2b}{2}$$

$$1.5 = b$$

$b = $ ....1.5............ ✓

[3]

[Total 6 marks]

(working in left margin)
$\begin{array}{r} 1.5 \\ 2\overline{)3\,0} \end{array}$

5 Solve the equation $(x + 2)(x - 4) = (x - 2)(x + 1)$.

Start by expanding the brackets on both sides of the equation.

$$(x \times x) + (x \times -4) + (2 \times x) + (2 \times -4) = (x \times x) + (x \times 1) + (-2 \times x) + (-2 \times 1)$$
$$x^2 \quad -4x + 2x - 8x = x^2 + x - 2x - 2$$

$$x^2 - 10x = x^2 - x - 2$$
$$x^2 - x^2 = -x + 10x - 2$$
$$= 9x - 2$$

$$x^2 - 2x - 8 = x^2 - x - 2$$
$$-8 + 2 = -x + 2x$$
$$-6 = x \qquad 80 \qquad x = -6$$

$x = $ ....

[Total 4 marks]

6 Solve the equation $6w^2 = 600$.

$$6w^2 = 600$$
$$w^2 = 100$$
$$w = \pm\sqrt{100}$$
$$w = \pm 10$$

$w = $ ....

[Total 3 marks]

Exam Practice Tip

It's a good idea to check your solution by substituting it back into the equation and checking that everything works out properly. It certainly beats sitting and twiddling your thumbs or counting sheep for the last few minutes of your exam. And if you have 'something' squared, don't forget the ± when you're solving it.

Score

18

25

Section Two — Algebra

Expressions, Formulas and Functions

1 $S = 4m^2 + 2.5n$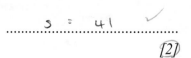

    a) Calculate the value of S when $m = 2$ and $n = 10$.

$$S = (4 \times \underline{2^{\text{th}}} \times \underline{2}) + (2.5 \times \underline{10})$$
$$S = \underline{16} + \underline{25}$$
$$S = \underline{41}$$

                                               $S = 41$ ✓

                                                          [2]

    b) Calculate the value of S when $m = 6.5$ and $n = 4$.

$$S = 4m^2 + 2.5n$$
$$= (4 \times 6.5 \times 6.5) + (2.5 \times 4)$$
$$= 169 + 10 = 179$$

                                              179 ✓

                                                            [2]

                                         [Total 4 marks]

2 Select the correct words from the box to complete the statements below.

a formula   an equation   an expression   a function

    a) $4q - 5$ is ......an......$expression$.... ✓

                                                            [1]

    b) $x^2 + 3x = 0$ is ....an....$equation$........... ✓

                                                            [1]

                                         [Total 2 marks]

3 The function machine below shows the function 'add 7 and divide by 5'.

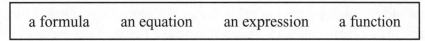

$x \longrightarrow \boxed{+7} \longrightarrow \boxed{\div 5} \longrightarrow y$

    a) Find the value of y when $x = 23$.

$$23 + 7 = 30$$
$$30 \div 5 = 6$$

                                          $y = $ ....6............ ✓

                                                              [1]

    b) Find the value of x when $y = 3$.

$$3 \times 5 = 15$$
$$15 - 7 = 8$$

                                          $x = $ ....8............ ✓

                                                              [2]

                                         [Total 3 marks]

Score:  9

9

Equations from Words and Diagrams

1   To convert kilometres into miles, Tasmin says that you divide
    the number of kilometres by 8 and multiply the answer by 5.

a) Write this rule as a formula.
   Use k to represent the number of kilometres and m to represent the number of miles.

   number of miles = (number of km $\div 8$) $\times 5$

   ~~8km~~

   $m = 8k \times 5$ $m = (k \div 8) \times 5$

                            $m = \frac{5k}{8}$

   .... $m = 8k \times 5$ ....

   [2]

b) Use your formula to convert 110 kilometres into miles.

   $m = 8(110) \times 5$

   $= 880 \times 5$ $m = \frac{5 \times 110}{8}$

   $= 4400$

   $m = 550 \div 8 = 68.75$

   .... 4400 .... miles

   [2]

   [Total 4 marks]

2   Nancy, Chetna and Norman are baking cakes for a cake stall. Chetna bakes twice as many
    cakes as Nancy and Norman bakes 12 more cakes than Chetna. They bake 72 cakes in total.

How many cakes does each person bake?

   N - N = 60

   Nancy = x

   Chetna = $2x$

   Norman = $(2x) + 12$

   $x + 2x + 2x + 12$

   $72 = 5x + 12$

   $-12 \qquad -12$

   $60 = 5x$

   $\frac{60}{5} = \frac{5x}{5}$

   Nancy $\leftarrow 12 = x$

   Chetna $\rightarrow 12 \times 2 = 24$

   Norman $\rightarrow 24 + 12 = 36$

   Nancy: .... 12 .... , Chetna: .... 24 .... , Norman: .... 36 ....

   [Total 4 marks]

3   Look at the rectangle below.

   Not drawn to scale

   $(x - 4)$ cm

   $(4x - 3)$ cm

a) Write a formula for P, the perimeter of the rectangle,
   in terms of x.

   $P = (4x - 3) + (x - 4) + (4x - 3) + (x - 4)$

   $= 10x - 14$

   $P = (4x - 3) + (x - 4)$

   $= (5x - 7)$

   $P = $ .... $5x - 7$ .... cm

   [2]

b) The perimeter of the rectangle is 36 cm. Find the value of x.

   $36 = 5x - 7$

   $+ \quad 7 \qquad\qquad +7$

   $\overline{}$

   $\frac{43}{5} = \frac{5x}{5}$

   $8.6 = x$

   $36 = 10x - 14$

   $+14 \qquad\qquad +14$

   $\overline{}$

   $\frac{50}{10} = \frac{10x}{10}$

   $5 = x$

   $x = $ 8.6 5

   [2]

   [Total 4 marks]

4 Jessica and Ricardo each think of a positive number.

a) Jessica squares her number, then subtracts 7. The result is 57.
What number is Jessica thinking of?

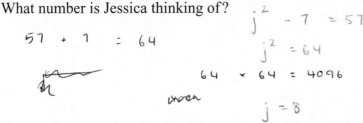

$57 + 7 = 64$

$j^2 - 7 = 57$

$j^2 = 64$

$64 \times 64 = 4096$

$j = 8$

.................4096................
[2]

b) Ricardo square roots his number, then adds 13. The result is 18.
What number is Ricardo thinking of?

$18 - 13 = 5$

$5 \times 5 = 25$

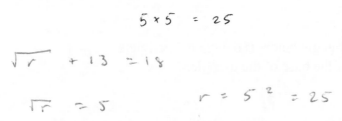

$\sqrt{r} + 13 = 18$

$\sqrt{r} = 5$

$r = 5^2 = 25$

.................25................
[2]

[Total 4 marks]

5 Find the length of one side of the equilateral triangle below.

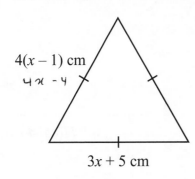

$4(x-1)$ cm
$4x - 4$

$3x + 5$ cm

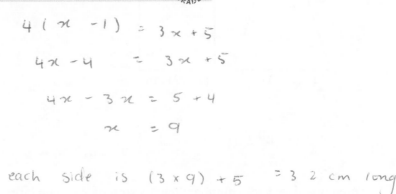

$4(x - 1) = 3x + 5$

$4x - 4 = 3x + 5$

$4x - 3x = 5 + 4$

$x = 9$

So each side is $(3 \times 9) + 5 = 32$ cm long

...................32................. cm
[Total 3 marks]

6 Peter, Cassie and Lisa are running a long-distance relay race.
Cassie takes 2 minutes longer than Peter to run her section of the race, and Lisa
is 4 minutes quicker than Peter. Their total time for the race is 43 minutes.

How long does each person take?

Hint: call Peter's time t, then
find expressions for Cassie and
Lisa's times in terms of t.

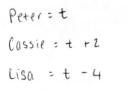

Peter = t
Cassie = t + 2
Lisa = t - 4

$t + t + 2 + t - 4$

$43 = 3t - 2$
$+ 2 \qquad + 2$
$45 = 3t$

$\dfrac{45}{3} = \dfrac{3t}{3}$

$15 = t$

Cassie
$15 + 2 = 17$

Lisa
$15 - 4 = 11$

Peter: ......15...... mins, Cassie: ......17...... mins, Lisa: ......11...... mins

[Total 3 marks]

Section Two — Algebra

7 At a football match between Redwood Rovers and Whitewater Wanderers, there were three times as many Redwood fans as Whitewater fans. The difference between the number of fans for each team was 7000.

How many fans were there in total?

Whitewater fans = f
Redwood fans = 3 × f = 3f

Difference = 3f - f = 2f, so 2f = 7000, so f = 3500

Total fans = 3f + f = 4f = 4 × 3500 = 14000

....

[Total 3 marks]

8 The perimeter of the isosceles triangle below is double the perimeter of the square. Find the length of the base of the triangle.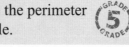

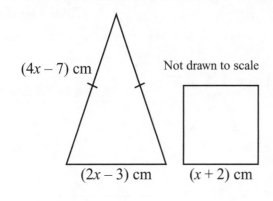

$(4x - 7)$ cm   Not drawn to scale

$(2x - 3)$ cm $(x + 2)$ cm

P△ = (4x - 7) + (4x - 7) + (2x - 3)
    = 10x - 17 cm

P□ = 4 × (x + 2) = 4x + 8 cm

So 10x - 17 = 2(4x + 8)

    10x - 17 = 8x + 16

    2x = 33

    x = 16.5 cm

.... cm

[Total 4 marks]

So base of △ is (2 × 16.5) - 3 = 30 cm

9 Three positive whole numbers have a sum of 25. The largest number is a multiple of 4, and the middle number is four times the smallest. What are the three numbers?

Call the largest number x and the smallest number y.

Then the middle number = ...4.... × ....y.... = ...4y...

So the sum of the three numbers = ...x.... + ...y.... + ...4y... = ... + ...5y...

So ....x.... + ...5y.... = 25

Now find x: multiples of 4 = ...4..., ...8..., ...12..., ...16..., ...20..., ...24...

25 − x needs to give a multiple of 5, so x = ...20...

Then ...20... + ...5y... = 25

    5y = 5

    y = 1

You could use trial and error to find numbers that work.

.... , ...4... , ...20...

[Total 3 marks]

Score: 7

32

Section Two — Algebra

Rearranging Formulas

1 The formula $v = u + at$ can be used to calculate the speed of a car.

a) Rearrange the formula to make u the subject.

$v = u + at$

$v - at = u$

.... $u = v - at$ ....
[1]

b) Rearrange the formula to make t the subject.

$v = u + at$

$v - u = at$

$t = \dfrac{v - u}{a}$

.... $t = \dfrac{v-u}{a}$ ....
[2]

[Total 3 marks]

2 Rearrange the formula $\dfrac{a+2}{3} = b - 1$ to make a the subject.

$a + 2 = 3b - 3$

$a = 3b - 5$

.... $a = 3b - 5$ ....
[Total 2 marks]

3 Rearrange the formula $x = y^2 - 7$ to make y the subject.

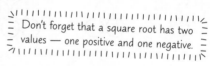

$x + 7 = y^2$

$y = \pm\sqrt{x + 7}$

....
[Total 2 marks]

4 Rearrange the formula $u = 2 + \dfrac{1}{w}$ to make w the subject.

$u = 2 + \dfrac{1}{w}$

$u - \underline{\ \ 2\ \ } = \dfrac{1}{w}$

$\underline{\ w\ }(\underline{\ u\ } - \underline{\ 2\ }) = \underline{\ 1\ }$

$\underline{\ w\ } = \dfrac{1}{u - 2}$

.... $w = \dfrac{1}{u-2}$ ....
[Total 3 marks]

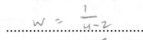

Score:
10

Section Two — Algebra

Sequences

1 Here are the first four terms of an arithmetic sequence. (GRADE 3)

[calculator crossed out icon]

4   12   20   28

a) Write down the next two terms in the sequence.

```
 '36
+   8
   44
```

...........36........ and ....44........
[1]

b) The 25th term in the sequence is 196. Write down the 23rd term in the sequence.

196 - 16 = 180

.........180......... *[1]*

c) Thomas says that the 12th term is 90. Explain why he is incorrect.

All terms in the sequence must be a multiple of 4. 90 isn't a multiple of 4, so it can't be the 12th term.

[2]

[Total 4 marks]

2 The patterns in the sequence below represent the first three triangle numbers. (GRADE 3)

(i) 1   (ii) 3   (iii) 6   a) Draw the next pattern in the sequence.

[1]

b) How many circles are in the tenth pattern in the sequence?
Give a reason for your answer.

1 + 2 + 3 + 4 + 5 + 6 + 7 + 8 + 9 + 10 = 55

The number of circles added increases by one each time.

[2]

[Total 3 marks]

3 To find the next term in the sequence below, you add together the two previous terms. (GRADE 4)

Fill in the gaps to complete the sequence.

| 3 | 4 | 7 | 11 | 18 | 29 |

[Total 2 marks]

4 The first four terms in a sequence are 2, 9, 16, 23, … **(4)**

a) Find the *n*th term of the sequence.

2ﾠﾠ9ﾠﾠ16ﾠﾠ23
ﾠﾠ7ﾠﾠﾠ7ﾠﾠﾠ7

The common difference is ..7.., so ..7..n is in the formula.

$$
\begin{array}{ccccc}
n = & 1 & 2 & 3 & 4 \\
7n = & 7 & 14 & 21 & 28
\end{array}
$$

↓ -5 ↓ -5 ↓ -5 ↓ -5 You have to subtract ..5.. to get to the term.

nth term = ..2.. ..9.. ..16.. ..23..

So the expression for the nth term is ..7n.. − ..5..

7n − 5

[2]

b) What is the 30th term of the sequence?

7n − 5
7(30) − 5 = 205

205

[1]

c) Is 55 a term in this sequence? Explain your answer.

$$55 = 7n - 5$$
$$+5 \qquad +5$$
$$\frac{60}{7} = \frac{7n}{7}$$
$$8.5714285\dot{} = n$$

No, because 'n' is not a whole number, so 55 is not in the sequence.

[2]

[Total 5 marks]

5 A quadratic sequence starts 2, 6, 12, 20, … **(4)**

4ﾠﾠ6ﾠﾠ8

Find the next term in the sequence.

20 + 10 = 30

Find the pattern in the differences between each pair of terms and use this to find the next term.

30

[Total 2 marks]

6 The *n*th term of a sequence is given by the rule $3n - 10$.
Two consecutive terms in the sequence have a sum of 1. **(5)**

What are the two terms?

The two consecutive terms will be the nth term and the (n + 1)th term. Use the rule for the nth term to find expressions for them.

3n − 10 and 3(n + 1) − 10 = 3n − 7

Sum = 3n − 10 + 3n − 7 = 6n − 17

So 6n − 17 = 1

6n = 18

n = 3

(3×3) − 10 = −1, (3×3) − 7 = 2

[Total 4 marks]

Score: ⎡ 11 ⎤

20

Section Two — Algebra

Inequalities

1   Write down the inequality shown on the number line below.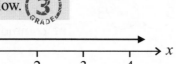

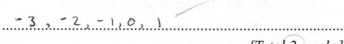

$x \geq -2$

[Total 1 mark]

2 n is an integer. List all the possible values of n that satisfy the inequality $-3 \leq n < 2$.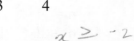

$-3, -2, -1, 0, 1$

[Total 2 marks]

3 p and q are integers. $p \leq 45$ and $q > 25$.

What is the largest possible value of $p - q$?

largest possible value of $p = 45$

smallest possible value of $q = 26$

largest possible value of $p - q = 45 - 26 = 19$

[Total 2 marks]

4   Solve the following inequalities.

a) $2p > 4$

$\dfrac{2p}{2} > \dfrac{4}{2}$

$p > 2$

$p > 2$

[1]

b) $4q - 5 < 23$

$+5 \quad +5$

$\dfrac{4q}{4} < \dfrac{28}{4}$

$q < 7$

$q < 7$

[2]

c) $4r - 2 \geq 6r + 5$

$4r \geq 6r + 7$

$4r - 2 \geq 6r + 5$

$-4r \qquad -4r$

$-2 \geq 2r + 5$

$-5 \qquad -5$

$-7 \geq 2r$

$\dfrac{-7}{2} \geq \dfrac{2r}{2}$

$-3.5 \geq r$

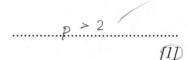

$-3.5 \geq r$

$r \leq -3.5$ *[2]*

[Total 5 marks]

Score: 5 / **10**

Section Two — Algebra

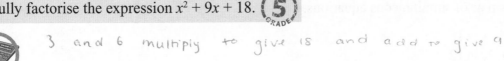

Quadratic Equations

1 Fully factorise the expression $x^2 + 9x + 18$.

   3 and 6 multiply to give 18 and add to give 9
so $x^2 + 9x + 18 = (x+3)(x+6)$

....

[Total 2 marks]

2 Fully factorise the expression $y^2 - 4y - 5$.

   $y^2 - 4y - 5 = (y + \underline{\ 1\ })(y - \underline{\ 5\ })$

....

[Total 2 marks]

3 Fully factorise the expression $x^2 + 4x - 32$.

....

[Total 2 marks]

4 The equation $x^2 - 9x + 20 = 0$ is an example of a quadratic equation.

a) Fully factorise the expression $x^2 - 9x + 20$.

....

[2]

b) Use your answer to part a) to solve the equation $x^2 - 9x + 20 = 0$.

$x = $ ....   or $x = $ ....

[1]

[Total 3 marks]

5 Solve the equation $x^2 + 4x - 12 = 0$.

$x = $ ....   or $x = $ ....

[Total 3 marks]

Score: ☐

12

Section Two — Algebra

Simultaneous Equations

1 Solve this pair of simultaneous equations.

$$4x + 3y = 16 \quad = 1$$
$$4x + 2y = ^-12 \quad = 2$$

$$0x + y = 4$$

$$y = 4$$

$$4x + 3(4) = 16$$
$$4x + 12 = 16$$
$$\quad -12 \quad -12$$
$$\frac{4x}{4} = \frac{4}{4}$$
$$x = 1$$

$$4(1) + 3(4)$$
$$4 + 12 = 16$$

$$4(1) + 2(4)$$
$$4 + 8 = 12$$

$$x = \underline{}1\checkmark \quad y = \underline{}4\checkmark$$

[Total 2 marks]

2 Solve this pair of simultaneous equations.

$$3x + 4y = 26$$
$$2x + 2y = 14$$

$$2x + 2y = 14 \xrightarrow{\times 2} 4x + \underline{4y} = \underline{28}$$
$$\quad - \quad 3x + 4y = ^-26$$
$$\qquad\qquad x = \underline{2}$$

$$\underline{6} + 4y = 26$$
$$4y = 26 - \underline{6} = \underline{20}$$
$$y = \underline{5}$$

$$x = \underline{2}\checkmark \quad y = \underline{5}\checkmark$$

[Total 3 marks]

3 Solve this pair of simultaneous equations.

$$x + 3y = 11$$
$$3x + y = 9$$
$$-2x + 2y = 2$$

$$x + 3y = 11 \quad ^1$$
$$3x + y = 9 \quad ^2$$

$$3x + 9y = 33$$
$$^-3x + y = 9$$
$$0x + 8y = 24$$

$$\frac{8y}{8} = \frac{24}{8}$$
$$y = 3$$

$$x + 3(3) = 11$$
$$x + 9 = 11$$
$$\quad -9 \quad -9$$
$$x = 2$$

$$2 + 3(3)$$
$$2 + 9 = 11$$

$$x = \underline{2}\checkmark \quad y = \underline{3}\checkmark$$

[Total 3 marks]

4 Solve this pair of simultaneous equations.

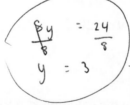
$$2x + 3y = 12 \xrightarrow{\times 5} 10x + 15y = 60$$
$$5x + 4y = 9 \xrightarrow{\times 2} 10x + 8y = 18$$

$$0x + 7y = 42$$

$$\frac{7y}{7} = \frac{42}{7}$$
$$y = 6$$

$$2x + 3(6) = 12$$
$$2x + 18 = 12$$
$$\quad -18 \quad -18$$
$$2x = -6$$

$$\begin{array}{r} 60 \\ -18 \\ \hline 42 \end{array}$$

$$\frac{2x}{2} = \frac{-6}{2}$$
$$x = -3$$

$$x = \underline{-3} \quad y = \underline{6}$$

[Total 4 marks]

Exam Practice Tip

Watch out — you might not actually be told they're simultaneous equations in the exam. But if you're told to solve two equations that look like the ones on this page (i.e. something x + something y = a number), you can be pretty sure you have to solve them simultaneously — you need to find values for x and y.

Score

12

Proof

1 For each statement below, write down an example to show that the statement is incorrect.

a) There are no factors of 48 between 15 and 20.

................16......?..

[1]

b) The sum of two square numbers is always odd.

............4....+....16....=....20..

[1]

c) All numbers that end in an 8 are multiples of either 4, 6 or 8.

....

[1]

[Total 3 marks]

2 Kit says "If I multiply any odd number by 3, the result is a multiple of 9."

Show that Kit is wrong.

$$9 \div 3 = 3$$

[Total 1 mark]

3 Prove that $(x + 2)^2 + (x - 2)^2 = 2(x^2 + 4)$ for all values of x.

 Start with the left-hand side and rearrange it until you get the right-hand side.

[Total 3 marks]

4 q is a whole number. Show that $2(18 + 3q) + 3(3 + q)$ is a multiple of 9.

[Total 3 marks]

Score:

10

Section Two — Algebra

Coordinates and Midpoints

1　Two points have been plotted on the grid below. They are labelled **A** and **B**.

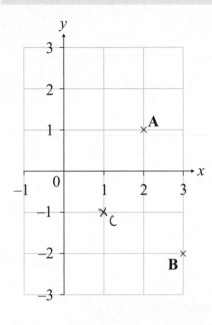

a) Give the coordinates of point **A**.

(.........2........, ...1........)
[1]

b) Give the coordinates of point **B**.

(.........3........, ...–2.......)
[1]

c) Point **C** has the coordinates $(1, -1)$.
Mark this point on the grid on the left using a cross (×) and label it **C**.
[1]

[Total 3 marks]

2　Points **A** and **B** have been plotted on the grid below.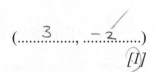

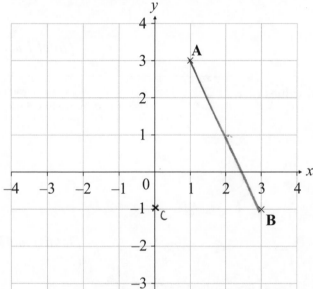

a) Write down the coordinates of the midpoint of the line segment **AB**.

$$\left(\frac{1 + \ \ 3\ \ }{2}, \frac{\ \ 3\ \ + \ \ -1\ \ }{2}\right) = (\ \ 2\ \ , \ \ 1\ \ )$$

(.........2........, ...1........)
[2]

b) Point **C** has the coordinates $(0, -1)$.
Given that line **AB** and line **CD** have the same midpoint, find the coordinates of point **D**.

x　distance $= 2 - 0 = 2$

y　distance $= 1 - -1 = 2$

So to get from the midpoint to point D, move up 2 and right 2

So point D is $(2+2, 1+2) = (4, 3)$

(..............., ...)
[2]

[Total 4 marks]

Score: 5

—　7

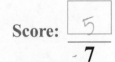

Straight-Line Graphs

1 Use the grid for the questions below.

   a) Draw and label the following lines.

$y = 3$

$x = -2$

$y = x$

[3]

b) What are the coordinates of the point where the lines $y = 3$ and $y = x$ meet?

(.........3......., ...3.........)

[1]

[Total 4 marks]

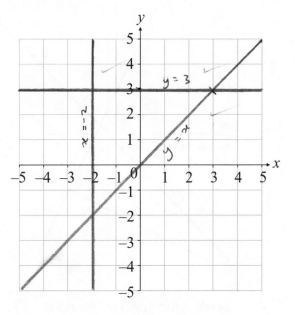

2 Answer each question below.

   a) Complete this table of values for the equation $y = 3x - 2$.

x	-2	-1	0	1	2
y	-8	-5	-2	1	4

[2]

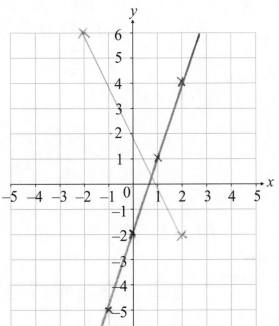

b) Use your table of values to plot the graph of $y = 3x - 2$ on the grid.

[2]

c) On the same grid, plot the graph of $y = -2x + 2$ from $x = -2$ to $x = 2$.

In the exam, you might not always be given a table of values to help you plot a graph — but it's a good idea to draw your own.

$y = -2x + 2$

$y = -2(2) + 2$

$= -4 + 2$

$= -2$

[3]

[Total 7 marks]

Section Three — Graphs

3 The graph below shows 3 lines — **a**, **b** and **c**.

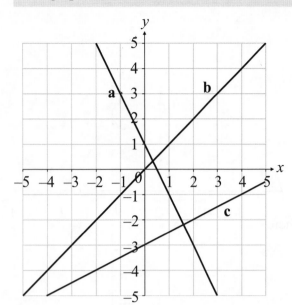

For each line, circle the correct gradient.

a) **a**

2 $-\frac{1}{2}$ $\frac{1}{2}$ $\boxed{-2}$

[1]

b) **b**

-1 $\frac{1}{3}$ 0 1

[1]

c) **c**

5 $\frac{1}{2}$ -3 -2

[1]

[Total 3 marks]

4 Look at the graph on the right.

Find the equation of the straight line.
Give your answer in the form $y = mx + c$.

$y = 3x + 1$

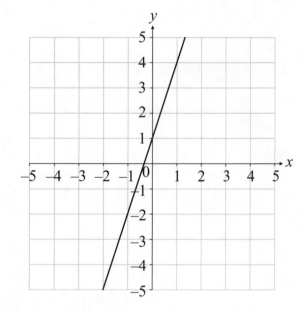

.........$y = 3x + 1$ ✓.............

[Total 3 marks]

5 Use the grid for the questions below.

a) Draw the line $y = x + 1$.

[2]

b) By drawing a second line, find the equation of
the line parallel to $y = x + 1$ which passes through
the point (2, 1).

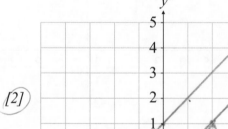

.........$y = x - 1$ ✓.............

[2]

[Total 4 marks]

Section Three — Graphs

6 $y = 4x - 3$ is the equation of a straight line.

Find the equation of the line parallel to $y = 4x - 3$ that passes through the point $(-1, 0)$.

$y = 4x + c$

$0 = 4(-1) + c$

$0 = -4 + c$ $y = 4x + 4$

$0 = -4 + 4$

.................$y = 4x + 4$.................

[Total 3 marks]

7 Line **L** passes through the points A $(1, -1)$ and B $(5, 7)$, as shown below.

a) Find the equation of line **L**.

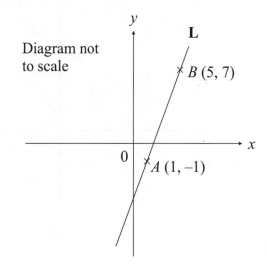

Diagram not to scale

$$\text{Gradient} = \frac{\text{change in } y}{\text{change in } x}$$

$$= \frac{-1 \quad - \quad 7}{1 \quad - \quad 5} = 2$$

Use point A to find c:

$$y = \overset{2}{\underset{m}{\ldots}} x + c$$

$$\overset{-1}{\underset{y}{\ldots}} = (\underset{2}{\ldots} \times \overset{1}{\underset{x}{\ldots}}) + c$$

$$c = \underset{-3}{\ldots}$$

So $y = \underset{2x - 3}{\ldots}$

....

[4]

b) A second line is drawn from the origin to the point $(4, 6)$.
    Is this line parallel to **L**? Give a reason for your answer.

$$m = \frac{6 - 0}{4 - 0} = 1.5$$

The gradient is different to Line L's gradient, so they can't be parallel.

[2]

[Total 6 marks]

Score: 21

30

Section Three — Graphs

Quadratic and Harder Graphs

1   The graph of $y = x^2 + 2x + c$ is shown below.

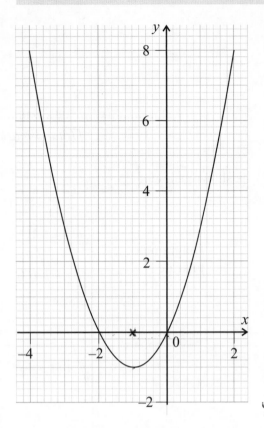

a) Circle the correct value of c.

(-2)   1   (0) -1 -4

[1]

b) Circle both solutions of $x^2 + 2x = 0$.

-1 (-2) ✓   3   (0) ✓   1

[1]

c) Write down the coordinates of the
turning point of the curve.

at $(-2, 0)$ and $(0, 0)$

x   coordinate   of
turning   point $= 1$

$y = 1^2 + 2(1) - 2$

$= 1 + 2 - 2$

$= 1$

($\frac{-1}{\dots}$, $\frac{-1}{\dots}$)
[1]

[Total 3 marks]

2   A table of values for $y = x^2 - 5$ is shown below.

x	-3	-2	-1	0	1	2
y	4	-1	-4	-5	-4	-1

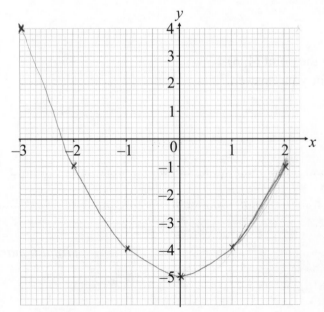

a) Draw the graph of $y = x^2 - 5$ on the grid.

[2]

b) Use your graph to estimate the negative
root of the equation $x^2 - 5 = 0$.
Give your answer to 1 decimal place.

This quadratic equation has two
roots. Make sure you choose
the one you're asked for.

$x = \dots$

[1]

[Total 3 marks]

Section Three — Graphs

3   Sketch the following graphs on the axes below.
Label the points where they intersect the axes.

a) $y = -x^2$

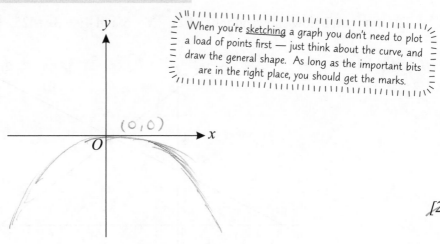

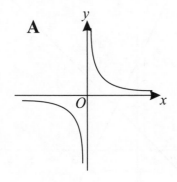

(0,0)

[2]

b) $y = x^3$

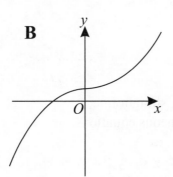

(0,0)

[2]

[Total 4 marks]

4   Sketches of different graphs are shown below.

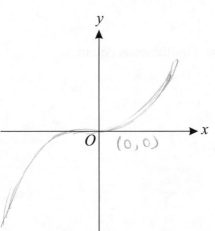

A   **B**   **C**

Write down the letter of the graph which shows $y = \dfrac{1}{x}$.

........A........

[Total 1 mark]

Score: 6

11

Section Three — Graphs

Solving Equations Using Graphs

1 The diagram below shows graphs of $2y - x = 5$ and $4y + 3x = 25$.

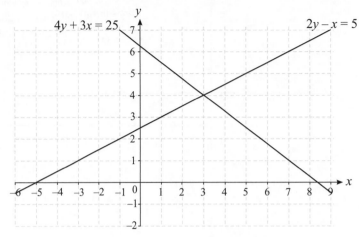

Use the diagram to solve these simultaneous equations:

$2y - x = 5$
$4y + 3x = 25$

$-x + 2y = 5$
$3x + 4y = 25$

$2(-20) - x = 5$
$-40 - x = 5$
$+40 \quad +40$

$4y - 2x = 10$
$4y + 3x = 25$

$3x - 6y = -15$
$3x + 4y = -25$

$x = 45$
$-x \quad -1$

$x = -45$ $\qquad$ $y = -20$ $\qquad$

$x = -15$
-1

$2y = -40$
$2 \quad 2$

$x = -45$

[Total 1 mark]

$x = -15$

$y = -20$

2 The diagram below shows the lines $y = x + 1$ and $y = 4 - 2x$.

a) Use the diagram to solve $x + 1 = 4 - 2x$

$x + 2x = 4 - 1$

$\dfrac{3x}{3} = \dfrac{3}{3}$

$x = 1$

$x = 1$ $\qquad$ ✓

[1]

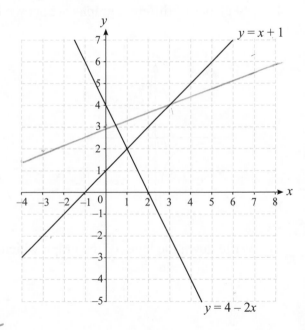

b) By drawing another straight line,
solve these simultaneous equations:

$y = x + 1 \quad -1 = x + y$
$3y = x + 9 \quad -9 = x - 3y$

$x + y = -1$
$x - 3y = -9$

$\dfrac{4y}{4} = \dfrac{8}{4}$

$x = 1$ $\qquad$ $y = 2$ $\qquad$

$y = 2$

[3]

[Total 4 marks]

$2 = x + 1$
$-1 \quad -1$

$1 = x$

Score: 2

5

Distance-Time Graphs

1 The distance-time graph below shows a 30 km running race between Selby and Tyrone.

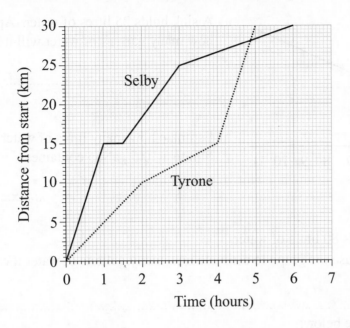

a) During the race Selby stops at a bench to get his breath back.
   After how many hours did he stop at the bench?

   ......1...... ✓ hour(s)
   [1]

b) Who won the race? How can you tell this from the graph?

   ...Tyrone, since the ~~4~~ line showed that he arrived earlier.✓............
   [1]

c) What was Selby's speed between 1.5 and 3 hours into the race? Give your answer to 2 d.p.

   $$speed = \frac{distance}{time} = \frac{10\ km}{1.5\ h}$$ ✓

   $$Speed = 6.66$$

   6.67
   ....6:7........... km/h
   [2]

d) During the race, one of the runners injured their leg.
   Which runner do you think was injured?
   What evidence is there on the graph to support your answer?

   ...Selby because it took her longer to finish the race than Tyrone...✓

   ..did...

   ....
   [2]

   [Total 6 marks]

Score: 5

6

Section Three — Graphs

Real-Life Graphs

1   Use the graph to help you answer the questions below.

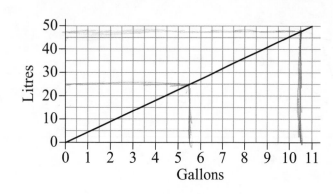

a) A sink holds 25 litres of water. Approximately how many gallons of water will it hold?

..........5.5....... gallons
[1]

b) Estimate how many litres of water would fit into a 10.5 gallon container.

.........40.7...... litres
47
[1]

c) Kirsty's paddling pool holds 80 litres of water.

(i)   Explain how she can use the graph to estimate how many gallons of water it can hold.

find 40 litres in gallons, then double the answer

.....She's able to see how much 80 litres would be in gallons.....

(ii)   Complete the sentence below:

17.6

Kirsty's paddling pool will hold ......160...... gallons of water.

[2]

$y = mx \text{ + }$ $80 = \frac{1}{2}x$ 40 litres ≈ 8.8 gallons
$y = \frac{1}{2}x$ $x = \frac{80}{0.5} = 160$ 80 litres ≈ 8.8 × 2 = 17.6   *[Total 4 marks]*

2   The table to the right shows how much petrol a car used on 3 journeys.

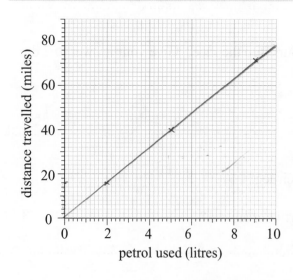

y Distance (miles)	16	40	72
x Petrol used (litres)	2	5	9

a) Using the table above, draw a conversion graph on the grid to the left.

[2]

b) Find the gradient of the line.

$\frac{y^2 - y^1}{x^2 - x^1}$

$\frac{72 - 16}{9 - 2} = \frac{56}{7} = 8$

..........8......

[2]

c) What does the gradient of the line represent?

....1 litre of petrol is used every 8 miles:..........................

distance travelled in miles per litre of petrol used

[1]

[Total 5 marks]

Score: 6

9

Section Three — Graphs

Ratios

1 Give the ratio $16:240$ in its simplest form.

$\frac{5}{6}$

$$16 \sqrt{\frac{15}{240}}$$
$$\underline{16\downarrow}$$
$$80$$
$$\underline{80}$$
$$0$$

$$\begin{array}{r} ^16 \\ \times\ 3 \\ \hline 48 \end{array}$$

$$\begin{array}{r} ^316 \\ \times\ 5 \\ \hline 80 \end{array}$$

.........$1 \therefore 15$.........

[Total 2 marks]

2 In a class of 26 children, 12 are boys and 14 are girls.

a) What is the ratio of boys to girls? Give your answer in its simplest form.

$$12 \therefore 14$$
$$6 \therefore 7$$

.........$6:7$.........

[1]

b) In another class, the ratio of boys to girls is $2:3$. There are 25 children in the class.
How many girls are there?

$$\times 5 \left(\begin{array}{ccc} 2 & : & 3 & \rightarrow & 5 \\ 10 & : & 15 & \rightarrow & 25 \end{array} \right) \times 5$$

.........15.........

[2]

[Total 3 marks]

3 Brian is making a fruit punch. He mixes apple juice,
pineapple juice and cherryade in the ratio $4:3:7$.

a) What fraction of the fruit punch is pineapple juice?

$$4 : 3 : 7 \Rightarrow 14$$
$$= 3/14$$

.........$3/14$.........

[1]

b) He makes 700 ml of fruit punch. What volume of each drink does he use?

$$\times 50 \left(\begin{array}{ccc} 4 & : & 3 & : & 7 & \longrightarrow & 14 \\ 200 & : & 150 & : & 350 & \rightarrow & 700 \end{array} \right) \times 50$$

Apple juice: ....200......... ml

Pineapple juice: ....150......... ml

Cherryade: ....350......... ml

[3]

[Total 4 marks]

4 The grid on the right shows two shapes, A and B.

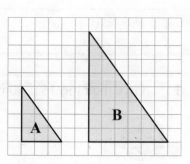

Give the following ratios in their simplest form.
a) Shortest side of shape A : shortest side of shape B

3 : 6

1 : 2

.... 1 : 2 ✓

[2]

b) Area of shape A : area of shape B

A = 12
B = 48   12 : 48
                1 : 4

.... 1:4 ✓

[3]

[Total 5 marks]

5 Last month a museum received £21 000 in donations. After taking off the cost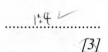
of monthly bills, the museum spent the remaining money on new exhibitions.

The ratio of bills to donations was 5 : 14. How much did they spend on new exhibitions?

x1500 (5 : 14
) x 1500
        0 : 21 000

7500 : 21000

21000 - 7500 = 13500

                                            13500
£ ....7.500...............

[Total 3 marks]
                                                2

6 Mr Appleseed's Supercompost is made by mixing soil, compost and grit in the ratio 4:3:1.
Soil costs £8 per 40 kg, compost costs £15 per 25 kg and grit costs £12 per 15 kg.

What is the total cost of materials for 16 kg of Mr Appleseed's Supercompost?

soil   = 8 : 40   ; 3.2 : 16
compost = 15 : 25   ;
grit   = 12 : 15

Hint: start by working out how much of each
material is needed for 16 kg of compost.

4 + 3 + 1 = 8 parts

soil = 4/8   compost = 3/8   grit = 1/8

16 kg contantains

4/8 x 16 = 8kg of soil → 8 ÷ 40 = £0.20/kg

3/8 x 16 = 6 kg of compost → 15 ÷ 25 = £0.60/kg

1/5 x 16 = 2kg of grit → 12 ÷ 15 = £0.80/kg

(8 x 0.2) + (6 x 0.6) + (2 x 0.8)

= £6.80

£ ......6.80..........

[Total 5 marks]

Section Four — Ratio, Proportion and Rates of Change

7 Bryn and Richard have just finished playing a game.
The ratio of Bryn's points to Richard's was 5:2.

a) On the axes below, draw a graph that could be used to work out Bryn's points if you know Richard's points.

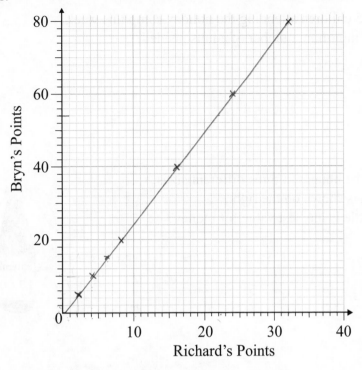

[2]

b) Richard scored 22 points. How many points did Bryn score?

.... 54 ~~55~~ points

[1]

[Total 3 marks]

8 Andy, Louise and Christine share a joint of beef in the ratio 3:6:7.
Christine gets 300 g more than Andy.

What is the total weight of the joint of beef?

A : L : C
3 : 6 : 7

Christine gets 7 parts & Andy gets 3 parts

So 300g ~~÷4~~ = 7 parts - 3 parts = 4 parts

1 part = 300g ÷ 4 = 75g

3 + 6 + 7 = 16 parts

total weight = 75 g × 16 = 1200 g

.... 1200 g

[Total 4 marks]

Score: 18

29

      Section Four — Ratio, Proportion and Rates of Change

Direct Proportion Problems

1 At a holiday camp there must be a minimum of 1 adult per 5 children.

 There are 95 children attending the holiday camp this week.
What is the minimum number of adults needed?

$95 \div 5 = 19$ ✓

$5 \overline{)9\cancel{4}5}$ 19

...........19 adults......... ✓

[Total 2 marks]

2 Brown sauce can be bought in three different sizes.
The price of each is shown on the right.
Which size of bottle is the best value for money?

250 ml £2.00   330 ml £2.75   525 ml £3.75

$\div 50 \left(\begin{array}{cc} 250 & : \pounds 2.00 \\ 5 & : 4p \end{array} \right)$ $\div 50 \left(\begin{array}{cc} 330 & : 2.75 \\ 5 & : 4.16p \end{array} \right) \div 66$ $\left(\begin{array}{cc} 525 & : 3.75 \\ 5 & : 3.571p \end{array} \right) \div 105$

..........525......... ml ✓

[Total 3 marks]

3 Joanna gets paid the same hourly rate whenever she works.
In the first week of July Joanna worked for 28 hours and got paid £231.
In each of the next 3 weeks of July, she worked for 25 hours.

How much will Joanna get paid in total for the 4 weeks she worked in July?

28 hrs $= \pounds 231$ $(\pounds 8.25/h)$ ✓

$8.25 \times 25 = \pounds 206.25$

$\cancel{231} \div 28 = \cancel{\pounds 8.25/h}$

$231 + (3 \times 25 \times 8.25) = \pounds 849.75$

£ 206.25.........

[Total 2 marks]

4 A football coach buys a bottle of water for each child in a football club.
All the bottles of water are the same price. There are 42 boys in the club.
He spends £52.50 on water for the boys. He spends £35 on water for the girls.

How many girls are there in the football club?

$52.50 \div 42 = \pounds 1.25$

$35 \div 1.25 = 28$

..........28......... ✓

[Total 2 marks]

Section Four — Ratio, Proportion and Rates of Change

5 Cat bakes 18 sponge cakes for an event in her village hall.
The recipe on the right will make 5 sponge cakes.

Ingredients

275 **g** flour (plain)
275 **g** butter
220 **g** sugar
5 eggs (medium)

a) Work out how much of each ingredient she used.

$18 \div 5 = 3.6$

flour - $275 \times 3.6 = 990 g$

butter - $275 \times 3.6 = 990 g$

sugar - $220 \times 3.6 = 792 g$

eggs - $5 \times 3.6 = 18$

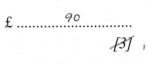

Flour: ....990............ g

Butter: ....990............ g

Sugar: ....792............ g

Eggs: ....18......eggs......

[3]

The total cost of the ingredients for 18 sponge cakes was £25.30.
She cut each cake into 10 slices and sold all the slices for 50p each.

b) How much profit did she make?

$18 \times 10 = 180$ slices

$50 \times 10 = 500p$

At 50p each of this £5.00 for each cake

$180 \times 50p = 9000p$

$\frac{9000}{5} \times 18 = £90$ = £90

£ ....90.............

[3]

$90 - 25.30 = £64.70$

[Total 6 marks]

6 The area of wallpaper (w m^2) required to cover all the walls in a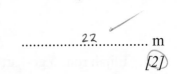
room is directly proportional to the perimeter (p m) of the room.
A kitchen has a perimeter of 17 m and 42.5 m^2 of wallpaper is needed.

a) A bedroom needs 55 m^2 of wallpaper. What is the perimeter of the room?

$$\begin{array}{cc} p & : & w \\ \div17 \left(\begin{array}{ccc} 17 & : & 42.5 \\ 1 & : & 2.5 \end{array} \right) \div 17 \end{array}$$

$$\times 22 \left(\begin{array}{ccc} 1 & : & 2.5 \\ 22 & : & 55 \end{array} \right) \times 22$$

...............22............... m

[2]

b) Sketch a graph on the axes below that shows the relationship between p and w.
Mark at least two points on your graph.

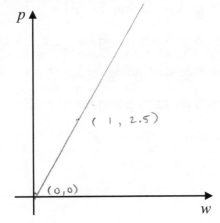

p

$(1, 2.5)$

$(0,0)$

w

[3]

[Total 5 marks]

Score: 16

20

Section Four — Ratio, Proportion and Rates of Change

Inverse Proportion Problems

1 It takes 12 people 3 hours to harvest the crop from a field.

Estimate how long it would take 4 people to harvest the crop.

12 × 3 = 36 hours for 1 person

36 ÷ 4 = 9 hours

.... 9 ✓ hours

[Total 2 marks]

2 A ship has enough food to cater for 250 people for 6 days.

a) For how many days can it cater for 300 people?

250 × 6 = 1506

1500 ÷ 300 = 5

.... 5 ✓ days

[2]

b) How many more people can it cater for on a 2-day cruise than on a 6-day cruise?

1500 250 × 6 = 1500

1500 ÷ 2 = 750

750 − 250 = 500

.... 500 ✓ people

[3]

[Total 5 marks]

3 Circle the **two** equations below that show that *f* is inversely proportional to *g*.

$$f = g^2 \qquad f = g + 5 \qquad \boxed{fg = 7} ✓ \qquad f = \frac{g}{5} \qquad \boxed{g = \frac{3}{f}} ✓$$

[Total 2 marks]

4 Elijah runs a go-kart track. It takes 12 litres of petrol to race 8 go-karts for 24 minutes.

6 go-karts used 18 litres of petrol. How many minutes did they race for?

1 litre of petrol will keep 8 go-karts going for $24 \div$ ...8...^{12}... = ...3...^{2}... minutes

18 litres of petrol will keep 8 go-karts going for ...3...^{2}... × 18 = ...54...^{36}... minutes

18 litres of petrol will keep 1 go-kart going for ...18...^{36}... × ...6...^{8}... = ...108...^{288}... minutes

18 litres of petrol will keep 6 go-karts going for ...108...^{288}... ÷ ...6... = ...18...^{48}... minutes

.... 18 minutes

[Total 4 marks]

Exam Practice Tip

Watch out for questions that involve a direct proportion as well as an inverse proportion — they can be tricky. If one does crop up on your exam, it's a good idea to write out your solution step-by-step. Then when you're checking your answers you can make sure each stage of your calculation makes sense.

Score
9
13

Section Four — Ratio, Proportion and Rates of Change

Percentages

1 Aled wants to buy a suitcase to take on holiday.

He sees a suitcase which was £18, but today it has 10% off.

a) How much money would he get off the suitcase if he bought it today?

10% of £18
= £18 ÷ 10 = £1.80

10% decrease
= 1 - 0.10 = 0.9

~~*10% of 18*~~

18 × 0.9 =   *0.10 × 18*

```
  0.10      7 0.9
×   18      × 1 8
  0 80        7 2
 1 0 0      0 90
 1 80      1.6.2
```

£ ......16.20........

[2]

He sees another suitcase which was £24, but the ticket says today the price is reduced by £6.

b) What is £6 as a percentage of £24?

6/24 ×100   *6/24 = 1/4*

1/4 × 100 = 100/4 = 25

........25 ✓.........%

[2]

[Total 4 marks]

```
    25
4)100
```

2 Work out 115% of 5200.

115% × 5200   *100% = 5200*   *0.115*
                       10% of 5200   *× 5200*
0.115 × 5200   *= 5200 ÷ 10 = 520*
                                            0 000
= 598   *5% of 5200 =*   *00 000*
                       520 ÷ 2 = 260   *0 23 000*
                                             0 575 000
                       115% = 5200 + 520 + 260   *0 59 8.000*
                       = 5980

..........5980......

[Total 2 marks]

3 Jane owns a fashion shop.

Jane sells a pair of jeans for £33.25 plus VAT at 20%.

How much does she sell the pair of jeans for?

20% of 33.25   *33.25 + 6.65 = 39.9*

0.20 × 33.25 = 6.65

£ ....39.90......

[Total 2 marks]

4 Franz always spends £2.40 a week on packs of football stickers.
The stickers normally cost 40p per pack but this week they are 40% cheaper.

How many more packs of stickers can he get this week than in a normal week?

£2.40 / week

~~*2.40 ÷ 0.96 = 2.5*~~

2.40 ÷ 0.40 = 6 packs normally

40% cheaper = 1 - 0.4 = 0.6   *15 - 6 = 9 ~~p~~ packs more.*

40% of £0.40 = 40p × 0.6 = 24p / pack this week

0.40 × 0.40 = £0.16 = 16p / pack after discount

240 ÷ 24 = 10 packs this week

~~*0.16 × 6 = 0.96 = 96p this week*~~

2.40 ÷ 0.16 = 15 packs this week

~~*10 - 6 = 4 more packs*~~

..........9.........

[Total 4 marks]

Section Four — Ratio, Proportion and Rates of Change

5 A school library contains 261 fiction books, 185 non-fiction books and 154 CDs.

100 books are borrowed from the library.
Now 41% of the items in the library are fiction books.
How many fiction books were borrowed from the library?

Hint: start by working out the total number of items remaining in the library.

261 + 185 + 154 = 600 −100 = 500 remaining items ✓

41% of 500

0.41 × 500 = 205 fiction books remaining ✓

261 − 205 = 56 fiction books were borrowed ✓

.............56.......... ✓

[Total 4 marks]

6 Oli and Ben each have a bank account that pays 8% simple interest per annum.
They each deposit an amount of money and don't pay in or take out any other money.

a) Oli deposits £2000 in the account. How much will be in the account after 3 years? **3**

'Per annum' just means 'per year'.

8% = 8 ÷ 100 = 0.08

8% of 2000 = 0.08 × 2000 = £160 / per year ✓

160 × 3 = £480   2000 + 480 = 2480

£ ....2480........

[2]

b) After the first year, Ben had £702 in his account. **4**
How much money did he originally put in the account?

0.08 × ____ = 702   £702 = 108%

~~0.08 ÷ 702~~

702 ÷ 108 = 6.50 = 1%

702 ÷ 0.08 = 8 775   6.50 × 100 = £650 = 100%

£ .....£650..........

At start of year =
£650 in his account

[3]

[Total 5 marks]

7 A pet rescue shelter houses cats and dogs. The ratio of cats : dogs is 3 : 7. **5**
40% of the cats are black and 50% of the dogs are black.

What percentage of the animals at the shelter are black?

3 : 7 means that 3 out of 10 = 30% of animals are cats

40% of 30% = 0.4 × 30% = 12% are black cats

100%   100% − 30% = 70% are dogs

50% of 70% = 0.5 × 70% = 35% are black dogs

So, 35% + 12% = 47% are black animals

.......................%

[Total 5 marks]

Score: 10

26

Compound Growth and Decay

1 Mrs Burdock borrows £750 to buy a sofa.
She is charged 6% compound interest per annum.

If Mrs Burdock doesn't pay back any of the money for 3 years, how much will she owe?
Give your answer to the nearest penny.

Multiplier = 1 + ...0.06.... = ...1.06...

After 3 years she will owe: .... × (.....................)⁻⁻⁻ = ...

After 1 year :   ~~10.000~~ × 1.06 = 795
                    750

After 2 years : 795 × 1.06 = 842.7   £ .....893.26.........

[Total 3 marks]

After 3 years : 842.7 × 1.06 = 893.262

2 The balance of a savings account, £B, is given by the formula $B = 5000 \times 1.02^t$
where t is the number of years since the account was opened.

a) What was the balance of the account when it was first opened?

£ ....5000............

[1]

b) How much is in the account after 7 years? Give your answer to the nearest penny.

5000×1.02^7 = 5143.428338

= 5143.43

£ .....5743.43.........

[2]

[Total 3 marks]

3 A car is travelling at 30 km/h. It starts to accelerate.
For every km it travels, its speed increases by 10%.

What will the car's speed be after it has travelled 5 km?
Give your answer to 3 significant figures.

Ben:

30×1.1^5 = 48.3153

= 48.3

.....48.3..... km/h
[Total 3 marks]

Score:

9

            Section Four — Ratio, Proportion and Rates of Change

Unit Conversions

1 Convert the following:

    a) 7.5 litres into millilitres.

.............7500............ ml

[1]

    b) 168 pounds into stones. (1 stone = 14 pounds)

$$14 \overline{\smash)168}^{12}$$

12 stones = 168 pounds

...........12.......... stone

[2]

[Total 3 marks]

2 Which of these measurements is the same as 0.87 metres?

    Circle around the correct answer.

      （ 87 mm ）   870 cm   870 mm   8.7 cm

[Total 1 mark]

3 Do the conversions below.

    a) Write 47 cm as a percentage of 2 m. ⟨200 cm⟩

    47% of 200

    0.47 × 200 = 94%.

    $\dfrac{47}{200} \times 100 = \dfrac{4700}{200}$

    $200\overline{\smash)4700} = 23.5$

.............23.5............ %

[2]

    b) Write 9 inches as a fraction of 3 feet. (1 foot = 12 inches)
       Give your answer in its simplest form. ⟨36 inches⟩

    $^9/_{36} = {}^3/_{12} = {}^1/_4$

.........¼.........

[2]

[Total 4 marks]

4 1 gallon = 8 pints. 9 litres ≈ 2 gallons.
    Approximately how many litres are there in 64 pints?

2 gallons = 16 pints   64 ÷ 16 = 4

9 litres ≈ 16 pints

9 × 4 = 36

36 litres ≈ 64 pints

.............36.......... litres

[Total 3 marks]

Section Four — Ratio, Proportion and Rates of Change

5 Nicole wants to post eight books to a friend in another country. Each book weighs 0.55 lb and postage is £0.50 per 100 g. (1 kg ≈ 2.2 lb)

How much will it cost her to post all the books to her friend?

$27 \div 0.55 = 4$ *(crossed out)*

$1 \div 4 = 0.25 \text{ kg} = 250 \text{ g}$ *(crossed out)*

$0.55 \text{ lb} = 250 \text{ g}$ *(crossed out)*

$£0.50 \times 2 = £1$ *(crossed out)*

$0.55 \text{ lb} \times 8 = 4.4 \text{ lb}$ *(crossed out)*

$0.55 \text{ lb} \times 8 = 4.4 \text{ lb}$

$2 \text{ kg} \approx 4.4 \text{ lb}$

$0.50 \times 2000 = £1000$

×2

£ ....1000.............

[Total 4 marks]

6 Naveed is tiling a rectangular section of wall measuring 1.6 m by 1.5 m. The tiles he is using are square, with a side length of 20 cm.

How many tiles does he need to cover the wall exactly? He can cut tiles in half if he needs to.

tiles = 20 cm

= 20 ÷ 100

= 0.2

0.2 m

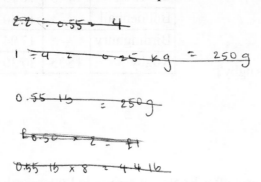

(crossed out: area)

Perimeter = 1.6 + 1.6 + 1.5 + 1.5 = 6.2 m

6.2 ÷ 0.2 = 31 tiles

.........31 tiles.........

[Total 3 marks]

7 A large cube has side lengths of 3 m. It is filled up with small cubes each with a side length of 60 mm.

 How many of the smaller cubes will fit inside the large cube?

60 mm = 6 cm = 6 ÷ 100 = 0.06 m

3 ÷ 0.06 = 50 small cubes

$$\begin{array}{r} 50 \\ 0.06\overline{)3\,00} \\ 30 \\ \hline 00 \end{array}$$

.........50 cubes.........

[Total 3 marks]

Score:

21

     Section Four — Ratio, Proportion and Rates of Change

Time Intervals

1 Part of the bus timetable from Coventry to Rugby is shown on the right.

Coventry	1445	1615	1745
Bubbenhall	–	1640	1810
Birdingbury	–	1704	–
Rugby	1535	1730	1840

a) Lisa arrives at Birdingbury bus stop at 16 58.
How long will she have to wait for the bus to Rugby?

The dashes on the timetable mean the bus doesn't stop.

4 :58 pm
5 :04 pm   +6

..........6......... minutes

[1]

The 16 15 bus from Coventry continues to Lutterworth after Rugby. It arrives in Lutterworth at 18 15.

b) If Anne catches this bus from Bubbenhall,
how long will it take her to get to Lutterworth? ②

18 ¹15
−16 40
‾‾‾‾‾‾
  1 75

1 hr 75 mins

= 2 hrs 15 mins

..........2........ hours ....15...... minutes

[2]

[Total 3 marks]

2 A cake has to be baked for 2¼ hours plus 10 minutes for every 100 g the cake weighs. ②

 Mary put a 400 g cake in the oven at 9.55 am.
What time should Mary take the cake out of the oven?

400g = 40 mins

2.25 hrs + 40 mins

= 2.65

= 3 hrs 5 mins

9 '55
+ 3 05
‾‾‾‾‾
12 6⁰

= 13 00 = 1 pm

..............1:00 pm.........

[Total 3 marks]

3 Isaac and Ultan spent 13 days building a model robot.
On the first 12 days they built from 4.30 pm till 7.15 pm and ②
on the last day they built for a total of 7 hours 10 minutes.

What is the total amount of time they spent building the robot?
Give your answer in hours and minutes.

⁶
₿ ¹15
− 4 30
‾‾‾‾‾‾
  2 85

:   3 hours 25 mins
+   7   10
‾‾‾‾‾‾‾‾‾‾‾‾‾‾‾‾‾‾
10 hours   35 mins

..............10...... hours ....35........ minutes

[Total 4 marks]

Score: ☐
10

Section Four — Ratio, Proportion and Rates of Change

Speed

1 John and Alan hired a van. Their receipt gave them information about how much time they spent travelling in the van, and how fast they went.

Travelling time: 1 hour 15 minutes
Average speed: 56 km/h

Calculate the distance that John and Alan travelled in the van.

Speed = d ÷ t

56 = d ÷ 75

d = 56 × 75 = 4200 km

.............4200............. km
[Total 2 marks]

2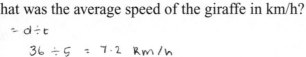

a) What was the average speed of the giraffe in km/h?

s = d ÷ t

36 ÷ 5 = 7.2 km/h

...........7.2........... km/h
[1]

b) What was the average speed of the giraffe in m/s?

S = d ÷ t

36 km = 36000 m

5 hours =

.... m/s
[2]

[Total 3 marks]

---

3 Adam has been caught speeding by a pair of cameras measuring average speed. The speed limit was 30 mph.

The cameras are 2.5 km apart. The time taken for his car to pass between them was 3 minutes.

a) What was Adam's average speed between the cameras? Give your answer to the nearest mph. (1 mile ≈ 1.6 km)

Speed = distance/time   2.5 km = 1.6 miles

1.6/3 = 0.53

.... mph
[3]

b) If Adam had been travelling within the speed limit, what is the minimum time it should have taken him to pass between the cameras? Give your answer to the nearest second.

.... s
[2]

[Total 5 marks]

Exam Practice Tip

Converting between different units of speed can be tricky. Just remember that speeds are made up of two measures — a distance and a time. If the units of each measure are changing e.g. mph to m/s, then you'll need to do two conversions, one for distance and one for time.

Score

10

Section Four — Ratio, Proportion and Rates of Change

Density and Pressure

1 The mass of a metal statue is 360 kg.
The density of the metal alloy from which it is made is 1800 kg/m³.

a) Calculate the volume of the statue.

.... m³

[2]

b) A different metal alloy is used to make a new statue.
The new statue has the same volume as the old one but has a mass of 220 kg.
Calculate the density of the new statue.

.... kg/m³

[2]

[Total 4 marks]

2 A metal alloy is made up of 120 g of metal A and 130 g of metal B.
Metal A has a density of 6 g/cm³ and metal B has a density of 5 g/cm³.

a) What is the total volume of metal used in the alloy?

> Hint: find the volume of each metal separately.

.... cm³

[3]

b) What is the density of the alloy? Give your answer to 1 decimal place.

.... g/cm³

[2]

[Total 5 marks]

3 Look at the cuboid on the right. Three of its faces are labelled A, B and C. The cuboid has a weight of 40 N.

Calculate the pressure, in N/m², that the cuboid exerts on horizontal ground when the cuboid is resting on face A.

80 cm   C   A   B   4 m   2 m

.... N/m²

[Total 3 marks]

Score:

12

Properties of 2D Shapes

1 Below are four shapes. **(1)** GRADE

A B C D

a) What is the mathematical name of shape B? ....

[1]

b) Which of these shapes is a rhombus? Write down the letter.

....

[1]

[Total 2 marks]

2 Below is a parallelogram. **(2)** GRADE

a) How many lines of symmetry does a parallelogram have?

....

[1]

b) What order of rotational symmetry does a parallelogram have?

....

[1]

[Total 2 marks]

3 An isosceles triangle has vertices A(1, 1), B(3, 7) and C(5, 1). **(2)** GRADE

Give the equation of its line of symmetry.

....

[Total 1 mark]

4 One of the angles in a rhombus is 62°. **(3)** GRADE

What are the sizes of its other three angles?

.............°, ...° and ....°

[Total 2 marks]

Score:

7

Congruent and Similar Shapes

1 Are triangles *ABC* and *DEF* congruent? Explain your answer.

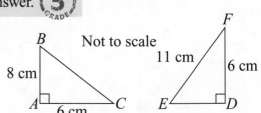

[Total 2 marks]

2 The shapes *ABCD* and *EFGH* are mathematically similar.

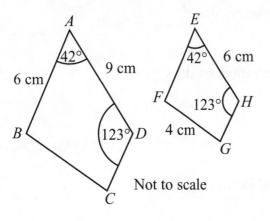

Not to scale

a) Find the length of *EF*.

.... cm
[2]

b) Find the length of *BC*.

.... cm
[1]

[Total 3 marks]

3 Triangles *ABC* and *DBE* are similar. *ABE* and *CBD* are both straight lines.

Find the missing values *x* and *y*.

Hint: You need to use the rule about vertically opposite angles to answer this question.

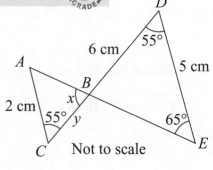

Not to scale

x = ...°

y = ... cm

[Total 3 marks]

Score:

8

The Four Transformations

1 Triangle **A** has been drawn on the grid below.

Reflect triangle **A** in the line $x = -1$. Label your image **B**.

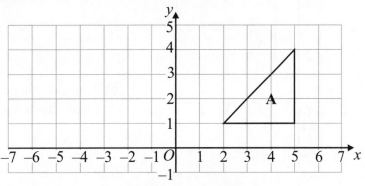

[Total 2 marks]

2 Shapes **F** and **G** have been drawn on the grid below.

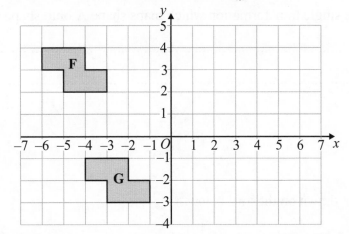

a) Write down the vector which describes the translation that maps **F** onto **G**.

....

[2]

b) Rotate shape **F** by 90° clockwise about the point (0, –2).
Label your image **H**.

[2]

[Total 4 marks]

3 On the grid enlarge the triangle by a scale factor of 3, centre (–4, 0).

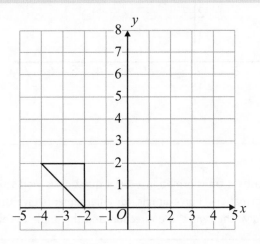

[Total 3 marks]

Section Five — Shapes and Area

4 The diagram below shows shape **A**.

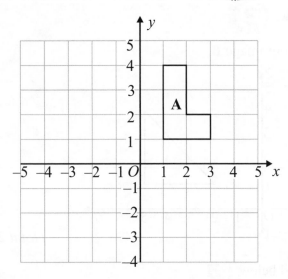

a) Rotate shape **A** by 180° about (–1, 0).
   Label your image **B**.

   [2]

b) Translate shape **B** by the vector $\begin{pmatrix} 2 \\ 4 \end{pmatrix}$.
   Label the image **C**.

   [1]

c) Describe fully the single transformation which maps shape **A** onto shape **C**.

   ....

   ....

   [3]

   [Total 6 marks]

5 Triangle **A** has been drawn on the grid below.

 Enlarge triangle **A** by a scale factor of $\frac{1}{2}$ with centre of enlargement (–6, 1).
Label your image **B**.

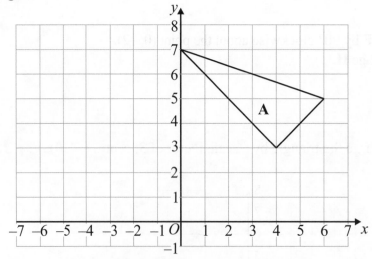

[Total 3 marks]

Exam Practice Tip

Make sure you give all the details when you describe a transformation — if a question is worth three marks then you'll probably need to give three bits of information. For example, for enlargements give the scale factor and the centre of enlargement, and for rotations give the centre, the direction and the angle of rotation.

Score

18

Section Five — Shapes and Area

Perimeter and Area

1 The shape below is drawn on a grid of centimetre squares.

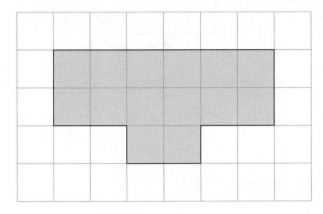

a) What is the perimeter of the shape?

.... cm
[1]

b) What is the area of the shape?

.... cm²
[1]

[Total 2 marks]

2 A shape is made up from an isosceles triangle and a trapezium.

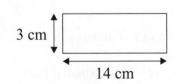

The area of the trapezium is 3 times as big as the area of the triangle.
a) Find the total area of the shape.

.... cm²
[3]

b) Find the height of the triangle.

.... cm
[2]

[Total 5 marks]

3 The diagram below shows a rectangle and a square.

3 cm ⟷ 14 cm   ? cm

Diagram not
accurately drawn

The ratio of the area of the rectangle to the area of the square is 6 : 7.
What is the area of the square?

.... cm²
[Total 2 marks]

Section Five — Shapes and Area

4 Malika is designing a logo to paint onto the front of her shop.

The background of the logo is in the shape shown
on the right, and she's going to paint it pink.
1 tin of pink paint covers 3 m².
Work out how many tins of paint she will need.

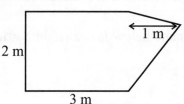

....

[Total 4 marks]

5 Lynn is designing a garden. The diagram shows her design.

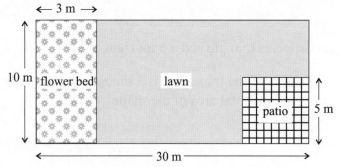

Diagram not
accurately drawn

Lynn's garden will be rectangular, with a rectangular flower bed at one end, and a
square patio at the other end. The rest of the space is taken up by a lawn.

Grass seed comes in boxes that cover 10 m² and cost £7 each.
How much will it cost Lynn to plant the lawn?

£ ....
[Total 3 marks]

6 The diagram below shows a rectangle with a right-angled triangle inside.

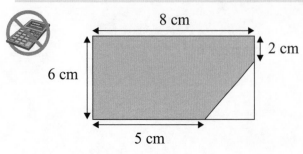

Calculate the area of the shaded part.

.... cm²
[Total 4 marks]

Score:

20

Section Five — Shapes and Area

Perimeter and Area — Circles

1   The diagram shows a circle with centre O.
    A, B and C are points on the circle.

a) What is the term given to the line AB? **(1)**

   ....
   [1]

b) Draw a straight line from the centre, O, to the edge of the circle. **(1)**
   What is the term given to this line?

   ....
   [1]

c) Draw a line connecting points B and C. **(2)**
   What is the term given to this line?

   ....
   [1]

d) At point A, draw a tangent to the circle. **(2)**

   [1]

   [Total 4 marks]

2   The radius of David's bicycle wheel is 0.25 m. **(3)**

a) Calculate the circumference of the wheel.
   Give your answer to 2 decimal places.

   .... m
   [2]

David's school is 500 m from his house.

b) Calculate how many complete turns the bicycle wheel
   makes when David rides to school from home.

   ....
   [2]

   [Total 4 marks]

3   A letter "O" is formed by cutting a circular section
    from the centre of a circular piece of card.

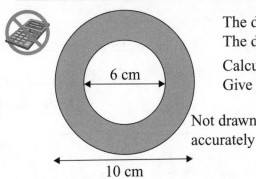

The diameter of the circle cut out is 6 cm.
The diameter of the circular card is 10 cm.

Calculate the exact area of the shaded region of the letter "O".
Give your answer in terms of π.

Not drawn
accurately

   .... cm²
   [Total 3 marks]

Section Five — Shapes and Area

4   For the semicircle on the right, find to 3 significant figures: **GRADE 4**

    a) its area,

14 mm

.... mm²

[2]

    b) its perimeter.

.... mm

[2]

[Total 4 marks]

5   The diagram below shows a square with a circle inside. **GRADE 4** The circle touches each of the four sides of the square.

Calculate the shaded area. Give your answer to 2 d.p.

8 m

.... m²

[Total 3 marks]

6   Look at the sector shown in the diagram below. **GRADE 5**

Diagram not accurately drawn

30°

6 cm

Find the perimeter and the area of the sector. Give your answers to 3 significant figures.

Don't forget to add the two radii to the arc length when finding the perimeter.

Perimeter = ... cm

Area = ... cm²

[Total 5 marks]

Exam Practice Tip

Don't mix up radius and diameter — it seems obvious, but lots of people muddle them up in exams. The radius of a circle is half of its diameter. Think carefully about which one you're being given, and which one you need for a formula. You won't be given the formulas in the exam, so make sure you know them off by heart.

Score

23

3D Shapes

1 Look at the table below.

Pentagon-based
pyramid

	Triangle-based pyramid	Square-based pyramid	Pentagon-based pyramid
Number of Faces	4	....	6
Number of Vertices	....	5	....
Number of Edges	6	....	10

a) Fill in the entries missing from the table.

[2]

b) Find a formula connecting E, the number of edges of a pyramid, and x, the number of sides of its base.

....
[2]

c) Use your formula to find the number of edges of a pyramid with an octagonal base.

....
[1]

[Total 5 marks]

2 Circle the net that makes a triangular prism.

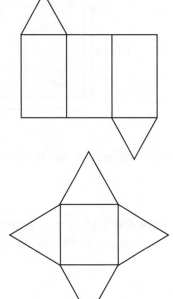

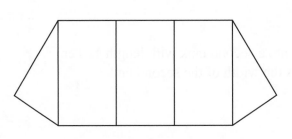

[Total 1 mark]

3 The diagram below shows the dimensions of a triangular prism.

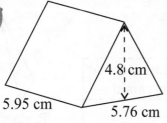

4.8 cm

5.95 cm   5.76 cm

Not drawn accurately

a) Estimate the volume of the triangular prism.

.... cm³

[2]

b) Does your estimate give an overestimate or
an underestimate of the volume of the prism?
Give reasons for your answer.

....

....

[1]

[Total 3 marks]

4 Volume of sphere = $\frac{4}{3}\pi r^3$.

Find the volume of the sphere on the right.
Give your answer to 3 significant figures.

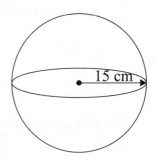

15 cm

.... cm³

[Total 2 marks]

5 The tank shown in the diagram below is completely filled with water.

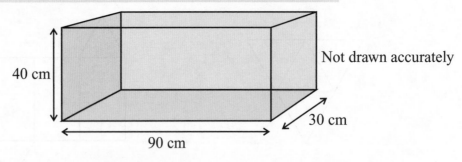

40 cm

Not drawn accurately

30 cm

90 cm

a) Calculate the volume of water in the tank.

.... cm³

[2]

b) The water from this tank is then poured into a second tank with length 120 cm.
The depth of the water is 18 cm. What is the width of the second tank?

.... cm

[2]

[Total 4 marks]

Section Five — Shapes and Area

6 The dimensions of a cube and a square-based pyramid are shown in the diagram below.

The side length of the cube is 7 cm. The side length of the pyramid's base is 2 cm and the slant height of the pyramid is 2 cm.

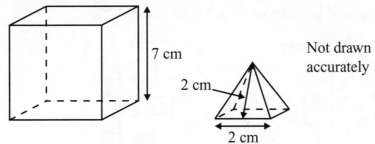

7 cm

2 cm

2 cm

Not drawn accurately

Find the ratio of the surface area of the cube to the surface area of the pyramid in the form $n:1$.

....

[Total 4 marks]

7 The diagram below shows a paddling pool with a radius of 100 cm.

Not drawn accurately

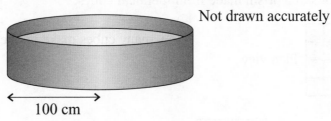

100 cm

The paddling pool is filled at a rate of 300 cm³ per second.
How long does it take to fill the pool to a depth of 40 cm?
Give your answer to the nearest minute.

Volume of water in paddling pool = ... × .... × h

= .... × 100² × ....

= ....

Time it will take to fill to depth of 40 cm = ... ÷ 300

= .... seconds

Convert to minutes = ... ÷ ....

= .... minutes

[Total 3 marks]

Exam Practice Tip

Be careful with the units in volume questions — you might be asked for a volume in m³, but given measurements in cm. Look out for this in rates of flow questions and don't let different units put you off. Just convert some of the units you're dealing with so they're all consistent — e.g. cm, cm³ and cm³ per second.

Score

22

Section Five — Shapes and Area

Projections

1 The diagram below shows a solid made from identical cubes.
The side elevation of the solid is drawn on the adjacent grid.

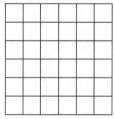

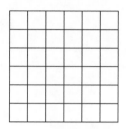

Side elevation

On the grids below, draw the front elevation and plan view of the solid.

Front elevation   Plan view

[Total 2 marks]

2 The diagram below shows the plan view, and the front and
side elevations of a prism made from identical cubes.

Plan view

How many cubes make up the shape?

Front elevation   Side elevation

....

[Total 2 marks]

3 The diagram below shows the plan, the front elevation and the side elevation of a prism.

Plan view

Draw a sketch of the solid prism on the grid below.

Front elevation   Side elevation

[Total 2 marks]

Score:

6

Section Five — Shapes and Area

Five Angle Rules

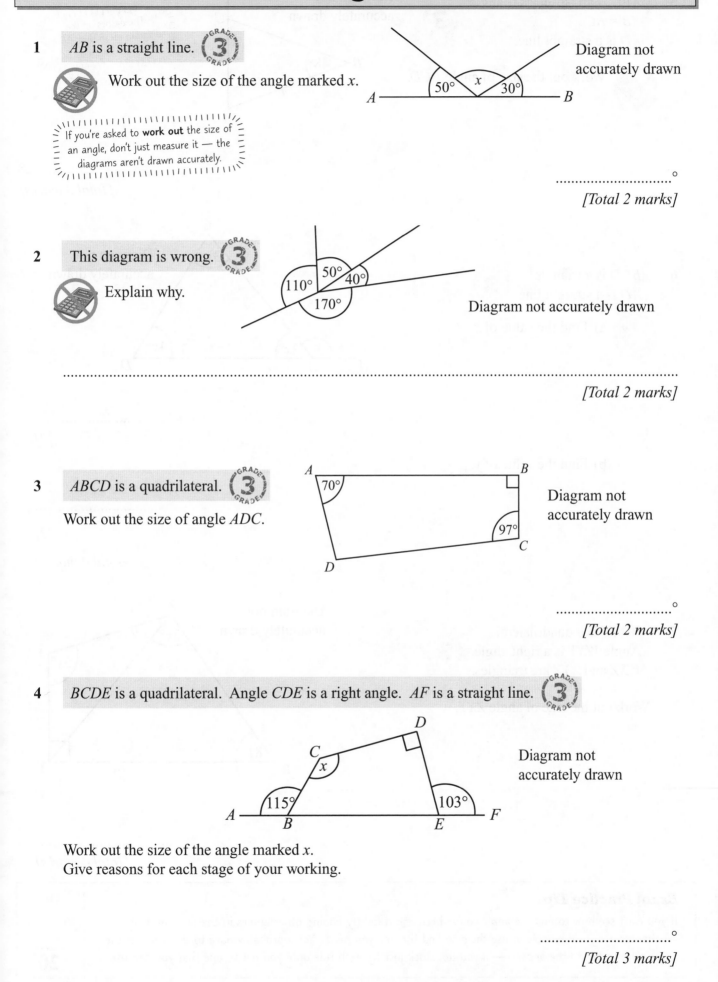

1 *AB* is a straight line. **GRADE 3**

Work out the size of the angle marked *x*.

\\\\\\\\\\\\\\\\\\\\\\\\\\\\\\\
If you're asked to **work out** the size of
an angle, don't just measure it — the
diagrams aren't drawn accurately.
/////////////////////////////

50° *x* 30°

A ———————————— *B*

Diagram not
accurately drawn

.............................°

[Total 2 marks]

2 This diagram is wrong. **GRADE 3**

Explain why.

110° 50° 40°
170°

Diagram not accurately drawn

....

[Total 2 marks]

3 *ABCD* is a quadrilateral. **GRADE 3**

Work out the size of angle *ADC*.

A 70° *B*

97° *C*

D

Diagram not
accurately drawn

.............................°

[Total 2 marks]

4 *BCDE* is a quadrilateral. Angle *CDE* is a right angle. *AF* is a straight line. **GRADE 3**

D

C
x

A ——— 115° ——————— 103° ——— *F*
      B   *E*

Diagram not
accurately drawn

Work out the size of the angle marked *x*.
Give reasons for each stage of your working.

.............................°

[Total 3 marks]

5 *ABC* is an isosceles triangle.
AB = *BC*.
AD is a straight line.

GRADE **3**

Diagram not
accurately drawn

The dashes on the diagram mean that *AB* is the same length as *BC*.

Work out the size of angle *BCD*.

..................................°

[Total 3 marks]

6 *BCD* is a triangle.
AD is a straight line.

GRADE **3**

Diagram not
accurately drawn

a) Find the value of *x*.

....

[2]

b) Find the value of *y*.

....

[2]

[Total 4 marks]

7 *WXYZ* is a quadrilateral.
Angle *WXY* is a right angle.
WXZ and *XYZ* are triangles.

GRADE **3**

Diagram not
accurately drawn

Work out the size of angle *ZXY*.

..................................°

[Total 4 marks]

Exam Practice Tip

If you can't see how to find the triangle you've been asked for, try finding other angles in the diagram first — chances are you'll be able to use them to find the one you need. You'll probably have to use a few of the angle rules to get to the answer — if you get stuck just try each rule until you get to one that you can use.

Score

20

Section Six — Angles and Geometry

Parallel Lines

1 Find the size of the angle marked *a*.
Give a reason for your answer.

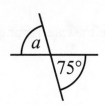

2 *BD* and *EF* are parallel straight lines.
AH is a straight line.

Work out the value of *x*.

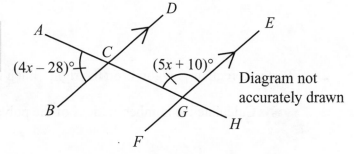

Diagram not accurately drawn

[Total 3 marks]

3 *BD* and *JH* are parallel straight lines.
AI and *EK* are straight lines.

Prove that triangle *EFG* is isosceles.

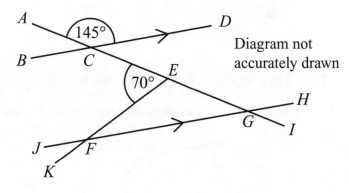

Diagram not accurately drawn

[Total 3 marks]

Score:

8

Section Six — Angles and Geometry

Angles in Polygons

1 Find the size of the exterior angle of a regular pentagon.

........................°

[Total 2 marks]

2 Part of a regular polygon is on the right. Each interior angle is 150°.

  Calculate the number of sides of the polygon.

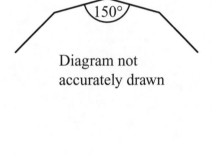

Diagram not
accurately drawn

....

[Total 3 marks]

3 The irregular polygon below has been divided into triangles as shown.

Use the triangles to show that the sum of the interior angles of the polygon is 900°.

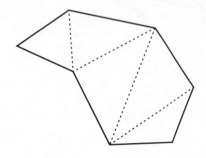

[Total 3 marks]

Score: ⬜

8

Section Six — Angles and Geometry

Triangle Construction

1 The diagram below is a sketch of triangle *ABC*.

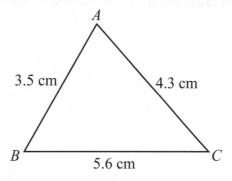

Diagram not
accurately drawn

Use a ruler and compasses to make an accurate drawing of triangle *ABC* in the space below.
You must show all your construction lines.

[Total 3 marks]

2 *EFG* is an isosceles triangle. Sides *EG* and *FG* are both 4.5 cm long.

Side *EF* has been drawn here.

E —————————— *F*

a) Complete the construction of triangle *EFG* by
drawing sides *EG* and *FG*.

[2]

b) Construct the bisector of angle *EGF*.

[2]

[Total 4 marks]

Score:

7

Section Six — Angles and Geometry

Loci and Construction

1 A dog is tied to a beam *AB* by a lead which allows it to run a maximum of 2 m from the beam.

 Find and shade the region on the diagram where the dog may run, using the scale shown.

Scale: 1 cm
represents 1 m

A —————————————— B

[Total 2 marks]

2 Using compasses and a ruler, draw an accurate perpendicular line from point R to line *ST*.

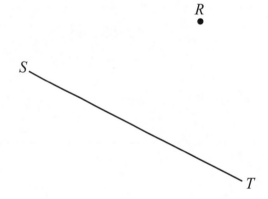

[Total 2 marks]

3 *ABC* is a triangle.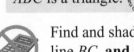

Find and shade the region inside the triangle which is **both** closer to the line *AB* than the line *BC*, **and** more than 6.5 cm from the point *C*.

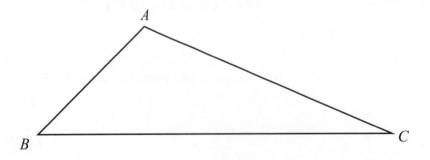

[Total 4 marks]

4 Hilary and Tony are deciding where they would like to put a pond in their garden. Hilary wants the centre of the pond to be exactly 1 m from the garden wall *BC*. Tony wants the centre of the pond to be exactly 2 m from the tree *F*.

 Accurately complete the plan of the garden below by placing crosses in the positions that Hilary and Tony would both be happy for the centre of the pond to be.

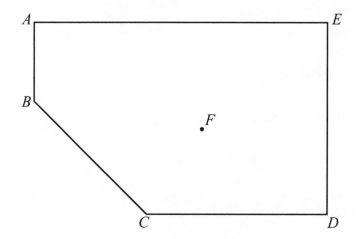

Scale: 1 cm represents 1 m

[Total 4 marks]

5 Triangle *XYZ* is shown below. It is rotated 180° clockwise about vertex *X* and then 90° clockwise about vertex *Z*.

Draw the locus of vertex *Y*.

Keep an eye on how vertex *Y* moves during each rotation.

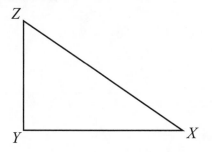

[Total 3 marks]

Score:

15

Section Six — Angles and Geometry

Bearings

1 Using the diagram below, find the three-figure bearing of Blackburn from Burnley.

N

⤊× Burnley

Blackburn ×

..........................°

[Total 1 mark]

2 Ruth cycles in a straight line from *V* to *U*.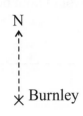

Find the bearing on which she travels to get from *V* to *U*.

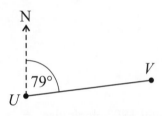

Diagram not
accurately drawn

N

79°

V

U

..........................°

[Total 2 marks]

3 Find the bearing of *B* from *A*.

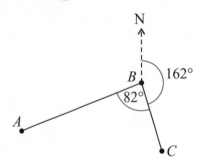

N

B ⋮ 162°

82°

A

C

Diagram not
accurately drawn

..........................°

[Total 2 marks]

Score: ⬚

5

Section Six — Angles and Geometry

Maps and Scale Drawings

1 Douglas drew a scale drawing of one of the rooms in his house. **(2)**

Cupboard

Dining table

Shelves

a) His dining table is 2 m long.
   What is the scale of this drawing?

   1 cm to .... m
   [1]

b) Work out the real distance from
   the dining table to the shelves.

   .... m
   [1]

c) Douglas wants to put a chair measuring 1 m × 1.5 m in the room
   so that there is a space of at least 0.5 m around it. Is this possible?
   Give a reason for your answer.

....

[2]

[Total 4 marks]

2 The instructions on a treasure map say "start at the cross and walk 400 metres on a **(3)**
bearing of 150°. Then walk 500 metres on a bearing of 090° to find the treasure."

Using a scale of 1 cm = 100 m,
accurately draw the path that
must be taken to find the
treasure on the map to the right.

Treasure Map

N

Start
×

Make sure you draw the north line
accurately for the second bearing.

[Total 3 marks]

Pythagoras' Theorem

1 The diagram shows a right-angled triangle *ABC*.
AC is 4 cm long. *BC* is 8 cm long.

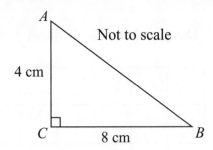

Calculate the length of *AB*.
Give your answer to 2 decimal places.

.... cm
[Total 3 marks]

2 An isosceles triangle has a base of 10 cm. Its other two sides are both 13 cm long.

Calculate the height of the triangle.

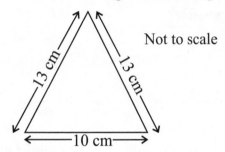

Think about whether you're trying to find
the hypotenuse or one of the shorter sides
before using Pythagoras' Theorem

.... cm
[Total 3 marks]

3 A rectangle has a height of 3 cm and a diagonal length of 5 cm.

Calculate the area of the rectangle.

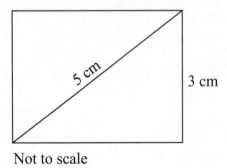

Not to scale

.... cm²
[Total 4 marks]

Score:

10

Section Six — Angles and Geometry

Trigonometry

1 The diagram shows triangle *ABC*.

Work out the length of *BC*.
Give your answer to 1 decimal place.

$$\cos x = \frac{\text{.............}}{\text{.............}}$$

$$\cos \text{.............} = \frac{\text{.............}}{\text{.............}}$$

.... × cos .... = ...

BC = ...

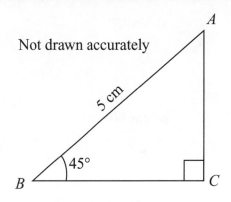

Not drawn accurately

.... cm

[Total 3 marks]

2 The diagram shows a right-angled triangle.

Work out the size of the angle marked *x*.
Give your answer to 1 decimal place.

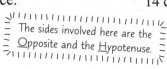
The sides involved here are the
Opposite and the Hypotenuse.

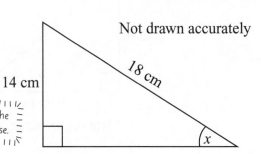

Not drawn accurately

14 cm

18 cm

x

............................°

[Total 3 marks]

3 The diagram shows a right-angled triangle.

Find *y*, giving your answer to 1 decimal place.

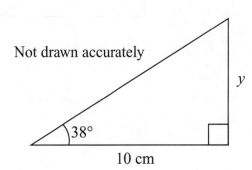

Not drawn accurately

y

38°

10 cm

.... cm

[Total 3 marks]

Section Six — Angles and Geometry

4 Circle the correct value of:

a) $\sin 45°$

$\dfrac{\sqrt{5}}{3}$ $\dfrac{1}{\sqrt{2}}$ $\dfrac{1}{2}$ $\dfrac{\sqrt{3}}{2}$ 2

[1]

b) $\cos 60°$

$\dfrac{4}{\sqrt{3}}$ 1 $\sqrt{2}$ $\dfrac{1}{2}$ $\dfrac{2}{3}$

[1]

[Total 2 marks]

5 Two right-angled triangles are shown below.

a) In this triangle, the longest side is exactly 1 m, and the lengths of the two shorter sides are given to 2 decimal places. Use this triangle to find $\sin 25°$ to 2 d.p.

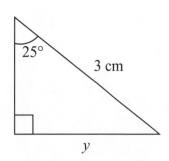

1 m

0.42 m

25°

0.91 m

Not drawn accurately

....

[1]

b) Use your answer to a) to work out the length of the side marked *y* to 1 decimal place.

25°

3 cm

y

Not drawn accurately

.... cm

[2]

[Total 3 marks]

Score: ☐

14

Vaectors

1 The diagram below shows a grid of unit squares.
Find, in terms of vectors **m** and **n**:

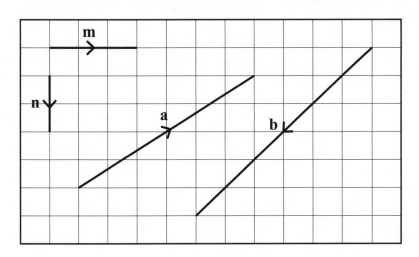

a) **a**

....
[1]

b) **b**

....
[1]

[Total 2 marks]

2 **a**, **b** and **c** are column vectors, where $\mathbf{a} = \begin{pmatrix} -3 \\ 5 \end{pmatrix}$, $\mathbf{b} = \begin{pmatrix} 5 \\ 4 \end{pmatrix}$ and $\mathbf{c} = \begin{pmatrix} -4 \\ -6 \end{pmatrix}$.

Calculate:
a) **a** – **b**

....
[1]

b) 4**b** – **c**

....
[1]

c) 2**a** + **b** + 3**c**

....
[1]

[Total 3 marks]

3 ABC is a triangle. $\overrightarrow{AB} = 2\mathbf{c}$ and $\overrightarrow{BC} = 2\mathbf{d}$. L is the midpoint of AC.

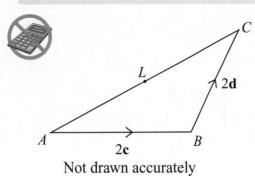

Not drawn accurately

Write in terms of $\mathbf{c}$ and $\mathbf{d}$:

a) $\overrightarrow{AC}$

....

[2]

b) $\overrightarrow{AL}$

$$\frac{1}{2} \times \underrightarrow{\text{..................}} = \frac{1}{2} \times \text{..................} + \frac{1}{2} \times \text{..................}$$

$$= \text{..................} + \text{..................}$$

....

[2]

c) $\overrightarrow{BL}$

....

[2]

[Total 6 marks]

4 ABC is a triangle where $\overrightarrow{AB} = 3\mathbf{a} + \mathbf{b}$ and $\overrightarrow{CB} = -6\mathbf{a} + 4\mathbf{b}$. P is the midpoint of BC.

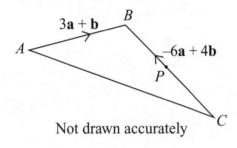

Not drawn accurately

a) Write $\overrightarrow{AC}$ in terms of $\mathbf{a}$ and $\mathbf{b}$.

....

[2]

b) Write $\overrightarrow{AP}$ in terms of $\mathbf{a}$ and $\mathbf{b}$.

....

[2]

[Total 4 marks]

Score: ▢

15

Section Six — Angles and Geometry

Probability Basics

1 Becky has a fair spinner with 8 equal sections as shown on the right.

a) Which colour is the spinner most likely to land on? **(1)**
   Circle your answer.

   Pink   Blue   Yellow

[1]

b) What is the probability of the spinner landing on a pink section? **(2)**

....

[1]

[Total 2 marks]

2 There are 10 counters in a bag. 4 of the counters are blue and the rest are red. **(2)**

One counter is picked out at random. On the scale below,
mark with an arrow (↓) the probability that a red counter is picked.

```
   |  |  |  |  |  |  |  |  |  |
   0              0.5              1
```

[Total 2 marks]

3 Steven records the positions of all the members of his **(2)**
football team. The table on the right shows his results.

A member of the team is chosen at random. What is the
probability they're a midfielder? Give your answer as a decimal.

Position	Frequency
Attacker	6
Midfielder	9
Defender	4
Goalkeeper	1

....

[Total 2 marks]

4 A game involves picking a ball at random out of **(3)**
one of the three bags shown on the right.
You win if the ball is black and lose if it's white.

Bag 1   Bag 2   Bag 3

Which bag would give you the greatest chance of winning?
Explain your answer.

....

....

[Total 2 marks]

Score: ☐

8

More Probability

1   Sarah has stripy, spotty and plain socks in her drawer.
She picks out a sock from the drawer at random.
This table shows the probability of her picking a spotty sock.

Sock	Stripy	Spotty	Plain
Probability	$2x$	0.25	x

    a)   What is the probability that she picks a sock that is **not** spotty?

....

[1]

    b)   What is the probability of her picking a stripy sock?

....

[2]

[Total 3 marks]

2   Katie decides to attend two new after-school activities. She can do one on Monday
and one on Thursday. Below are lists of the activities she could do on these days.

Monday
Hockey
Orchestra
Drama

Thursday
Netball
Choir
Orienteering

a) List all nine possible combinations of two activities Katie could try in one week.

[2]

Katie randomly picks an activity to do on each day.
Use your answer to part a) to find:

b) the probability that she does hockey on Monday and netball on Thursday,

....

[1]

c) the probability that she does drama on Monday.

....

[1]

[Total 4 marks]

3 Sammi has 3 pieces of homework, English (E), History (H) and Maths (M). She has to do all 3 pieces tonight but she can do them in any order.

a) List all the different orders in which she could do her pieces of homework.

....

....

[2]

Sammi randomly chooses the order in which to do her pieces of homework.

b) Use your answer to part a) to find the probability that she does her Maths homework before her English homework. Give your answer as a fraction in its simplest form.

....

[1]

[Total 3 marks]

4 Alvar has a fair 6-sided dice and a set of five cards numbered 2, 4, 6, 8 and 10. He rolls the dice and chooses a card at random. Alvar adds the number on the dice to the number on the card to calculate his total score.

a) Complete the table below to show all of the possible scores.

Cards

		2	4	6	8	10
	1					
	2					12
	3				11	13
Dice	**4**			10	12	14
	5		9	11	13	15
	6	8	10	12	14	16

[2]

b) Find the probability that Alvar will score 12 or more. Give your answer as a fraction in its simplest form.

....

[2]

c) Alvar says "All the cards are even numbers so I am more likely to get an even-numbered score than an odd-numbered score." Is he correct? Explain your answer.

....

....

[2]

[Total 6 marks]

Exam Practice Tip

When listing all the possible outcomes, it's easy to either miss some out or repeat yourself — so try to go through them in a sensible way. If each outcome is equally likely, you can find the probability of something by counting how many outcomes fit what you're being asked and dividing that by the total number of outcomes.

Score

16

Probability Experiments

1   The probability of a train arriving in Udderston on time is 0.64.

Hester will get the train to Udderston 200 times this year.
Estimate the number of times Hester will arrive in Udderston on time.

....

[Total 1 mark]

2   Georgie has a biased 5-sided spinner numbered 1-5.
The table below shows the probabilities of the spinner landing on numbers 1-4.

Number	1	2	3	4	5
Probability	0.3	0.15	0.2	0.25	

She spins the spinner 100 times. Estimate the number of times it will land on 5.

....

[Total 2 marks]

3   Suda has a 6-sided dice. The sides are numbered 1 to 6.
Suda rolls the dice 50 times. Her results are shown in the table below.

Number	1	2	3	4	5	6
Frequency	16	6	12	7	3	6
Relative frequency						

a) Complete the table above.

[2]

b) Suda says, "The dice has 6 sides so the probability of it landing on a 1 is $\frac{1}{6}$."
Criticise Suda's statement.

....

....

[2]

Suda rolls the dice another 200 times and records her results.

c) Which set of results will give more reliable estimates for the probabilities of the dice landing on each number, the first set or the second? Explain your answer.

....

....

[1]

[Total 5 marks]

Section Seven — Probability and Statistics

4   Danielle flipped a coin 100 times, and predicted the outcome before each flip.
    She predicted it would land showing heads 47 times and got 25 correct.
    Of the times she predicted it would land on tails, she got 26 correct.

a) Complete the frequency tree below to show these results.

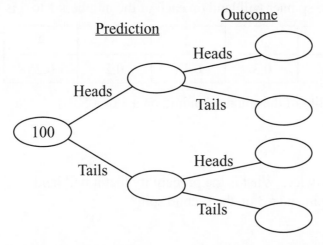

[2]

b) Work out the relative frequency of Danielle predicting the outcome correctly.

....

[2]

[Total 4 marks]

5   John throws a ball at a target using his left and right hands.
    His results are shown in the table on the right.

	Left Hand	Right Hand
Throws	20	100
Hit target	12	30

a) Estimate the probability that John will hit the
   target with his next throw if he uses his left hand.

....

[2]

b) John uses his results to estimate the probabilities of him hitting the target using each hand.
   Explain which of his estimated probabilities will be more reliable.

....

....

[1]

[Total 3 marks]

Score:

15

The AND/OR Rules

1   A biased 5-sided spinner is numbered 1-5.

The probability that the spinner will land on each of the numbers 1 to 5 is given in this table.

Number	1	2	3	4	5
Probability	0.3	0.15	0.2	0.25	0.1

a) What is the probability of the spinner landing on a 4 or a 5?

....

[2]

b) The spinner is spun twice. What is the probability that it will land
on a 1 on the first spin and a 3 on the second spin?

....

[2]

[Total 4 marks]

2   Shaun is playing the game 'hook-a-duck'.
The probability that he wins a prize is 0.3.

a) What is the probability that he does not win a prize?

....

[1]

b) If he plays two games, what is the probability that he doesn't win a prize in either game?

....

[2]

[Total 3 marks]

3   Alisha and Anton are often late for dance class.
The probability that Alisha is late is 0.9. The probability that Anton is late is 0.8.

What is the probability that at least 1 of them is late to the next dance class?

P(at least 1 is late) = 1 − P(neither is late)

P(Alisha isn't late) = 1 − .... = ...   P(Anton isn't late) = 1 − .... = ...

P(neither is late) = ... × .... = ...

P(at least 1 is late) = 1 − .... = ...

....

[Total 4 marks]

Score: [ ]

11

Tree Diagrams

1   Jo and Heather are meeting for coffee.
The probability that Jo will wear burgundy trousers is 0.4.
The probability that Heather will wear burgundy trousers is 0.25.

a) Complete the tree diagram below.

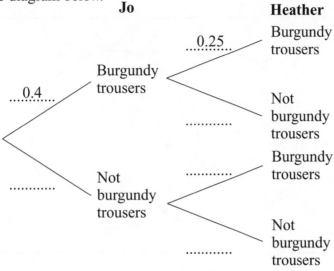

[2]

b) What is the probability that neither of them wear burgundy trousers?

....

[2]

[Total 4 marks]

2   A couple are both carriers of a gene that causes a disease.
If they have a child, the probability that the child will carry the gene is 0.25.

a) The couple have two children. Draw a tree diagram to show the probabilities
of each child carrying or not carrying the gene.

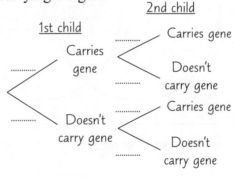

[2]

b) Find the probability that both children will carry the gene.

....

[2]

[Total 4 marks]

Score:

8

Sets and Venn Diagrams

1   The Venn diagram below represents 60 elements.
    39 elements are in set A. 45 elements are in set B.
    21 elements are in set B but not in set A.

Complete the diagram.

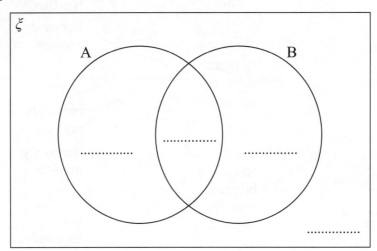

[Total 3 marks]

2   100 Year 7 students were asked if they like apples (A) or bananas (B).
    70 like apples, 40 like bananas and 20 like apples and bananas.

   It's a good idea to start by filling in the intersection.

a) Complete the Venn diagram below showing this information.

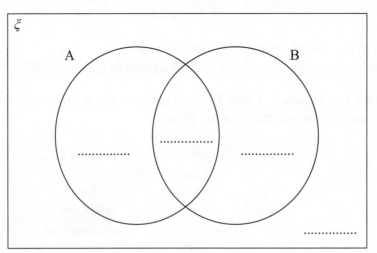

[3]

b) One of the students is selected at random.
    Find P(A ∪ B).

....

[2]

[Total 5 marks]

Score:

8

Sampling and Data Collection

1 Faye is investigating how many chocolate bars teenagers buy each week.
She is going to collect data by asking her teenage friends how many they buy.

a) Design a table that Faye could use to record her data. **(3)**

[2]

b) Comment on whether she can use her results to draw conclusions about teenagers in the UK. **(4)**

....

....

....

[2]

[Total 4 marks]

2 Mario asked 50 people at a football match how they travelled there. He found
that 22 of them travelled by car. There were 5000 people at the match altogether. **(4)**

a) Use the information above to estimate the number of people who travelled to the match by car.

....

Remember, to get reliable estimates, a sample
needs to fairly represent the population.

[3]

b) Comment on the reliability of your estimate in part a).

....

....

....

[1]

[Total 4 marks]

Score:

8

Mean, Median, Mode and Range

1 One evening Preya makes 10 phone calls. When the bill comes it shows how long each call was, in minutes. The call lengths are listed below.

| 10 | 12 | 25 | 3 | 37 | 13 | 12 | 18 | 41 | 33 |

a) Work out the median length of Preya's calls.

.... minutes

[2]

b) Calculate the mean phone call length.
   Give your answer to the nearest minute.

.... minutes

[2]

c) What is the range?

.... minutes

[1]

[Total 5 marks]

2 Sam thinks of three different whole numbers.

The numbers have a range of 6 and a mean of 4.
What are the three numbers?

.................., ..., ...

[Total 2 marks]

3 Lee has 6 pygmy goats. Their weights, in kg, are listed below.

| 32 | 23 | 31 | 28 | 36 | 26 |

a) Which three weights from the list above would have a range which is half the value of the median of the three weights? Also write the range and median of the three weights.

.................., ..., ...

range = ..., median = ...

[2]

b) Two of the goats wander off and don't return. The mean weight of the herd is now 27.25 kg.
   Find the weights of the two goats who wandered off.

.... kg and .... kg

[3]

[Total 5 marks]

Section Seven — Probability and Statistics

4 A bakery records the number of cookies it sells each day for ten days.
The mean number is 17 and the median number is 15.

The next day the bakery sells 18 cookies.

a) Is the mean number sold over all eleven days higher than 17? Explain your answer.

....
[1]

b) Is the median number sold over all eleven days higher than 15? Explain your answer.

....
[1]

[Total 2 marks]

5 25 people were asked how many holidays they went on last year.
The vertical line graph below shows the results.

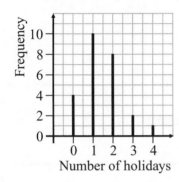

a) Write down the modal number of holidays.

....
[1]

b) Find the median number of holidays.

....

Imagine the data in a list – 0, 0, 0, 0, 1... Find the position
of the median and count up through the bars till you get there.

[2]

[Total 3 marks]

6 This data shows the amount of rainfall in mm that
fell on an island during a 12-day period in June.

a) Work out the range of the rainfall and comment on
this value as a measure of the spread of the data.

0	8
1	7 9
2	3 6 9
3	0 1 4 7 8
6	3

Key
0 | 8 means
8 mm of rain

....

....
[3]

In November the median amount of rainfall was 22 mm and the range was 20 mm.

b) Compare the rainfall in June with the rainfall in November.

....

....

....

Don't be put off by the way the data is displayed —
look at the key to work out how to read off the values.

[3]

[Total 6 marks]

Exam Practice Tip

You've been treated to two pages on averages and range, because as well as knowing the basics, you need
to be able to apply your knowledge to trickier questions. When you're comparing data sets using averages
and range, remember to explain what the values mean for those particular data sets.

Score

23

Simple Charts and Graphs

1 The dual bar chart below shows the number of cups of different hot drinks sold in a cafe last Saturday and Sunday.

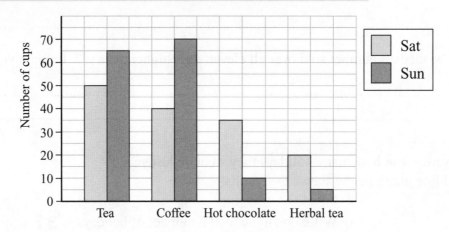

a) How many more cups of coffee than tea were sold on Sunday?

....

[1]

b) Which drink did the cafe sell 25 more cups of on Saturday than Sunday?

....

[1]

c) On which day were most hot drinks sold in total?

....

[2]

d) What fraction of the cups of herbal tea were sold on Saturday?
Give your answer in its simplest form.

....

[1]

[Total 5 marks]

2 This table shows some information about the favourite sports of some students.

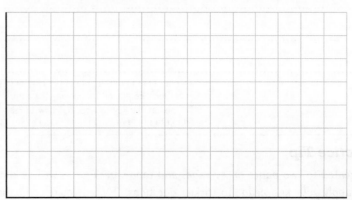

Sport	Students
Football	14
Swimming	5
Athletics	9
Netball	1
Hockey	6

Show this information as a bar chart on the grid below.

[Total 4 marks]

3 This stem and leaf diagram shows the number of newspapers a shop sold on each day in June.

Key: 0 | 5 means 5 newspapers

```
0 | 0 0 0 2 2 5 8 9
1 | 1 1 1 2 3 4 5 7 9
2 | 0 0 5 6 6 7 7 7 8
3 | 0 3 5 9
```

a) On how many days did the shop not sell any newspapers?

.... days
[1]

b) On what fraction of the days did the shop sell more than 30 newspapers?
Give your answer in its simplest form.

....
[1]

[Total 2 marks]

4 This pictogram shows the number of eggs laid by some chickens at a farm on Monday, Tuesday, Wednesday and Friday.

Monday	◯ ◯ ◯ ◔	◯ = 8 eggs
Tuesday	◯ ◯ ◖	
Wednesday	◯ ◯ ◯ ◯ ◯	
Thursday		
Friday	◯ ◯ ◯ ◯ ◯	

a) How many eggs were laid on Monday?

....
[1]

b) How many more eggs were laid on Wednesday than Tuesday?

....
[1]

c) 24 eggs were laid on Thursday. Show this information on the pictogram.

[1]

The farmer is going to use 40% of the eggs laid on Friday to make sponge cakes.

d) How many eggs is this?

....
[2]

[Total 5 marks]

5 This pictogram shows the number of jars of jam sold in a campsite shop in one month.

The key is missing from the pictogram.
The shop sold 100 jars of jam altogether.
How many jars of raspberry jam did it sell?

Strawberry Jam	🍓 🍓 🍓 🍓
Blackberry Jam	🍓 🍓 ◗
Raspberry Jam	🍓 🍓 🍓 ◗

....
[Total 4 marks]

Section Seven — Probability and Statistics

6 50 people were asked if they've ever been skiing. The table below shows the results.

	Have been skiing	Have not been skiing
Male	15	20
Female	5	10

a) Write down the number of males who have been skiing to the number of males who have not been skiing as a ratio in its simplest form.

....

[1]

b) What percentage of all the people asked have been skiing?

.... %

[2]

[Total 3 marks]

7 The numbers of swallows seen in Bluebell Wood over three years are shown in the table.

Year	2010				2011				2012			
Month	Jan	Apr	Jul	Oct	Jan	Apr	Jul	Oct	Jan	Apr	Jul	Oct
No. of swallows	0	60	44	13	0	57	36	10	0	56	34	6

This line graph has been drawn to show the data.

a) Write down one reason why this graph is misleading.

....

....

....

[1]

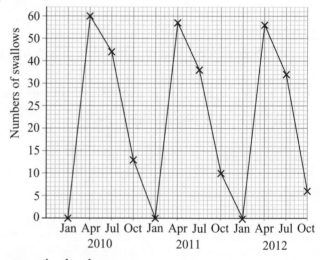

b) Use the table above to describe the repeating pattern in the data.

....

....

....

[1]

[Total 2 marks]

Score:

25

Pie Charts

1   A survey was carried out at a leisure centre to find out which sport people prefer to do. The results are shown in the pie chart.

a) What fraction of people prefer to do fitness training?

....
[1]

60 people said they prefer to play football.

b) How many people prefer to play badminton?

....
[2]

[Total 3 marks]

2   A survey was carried out in a local cinema to find out which flavour of popcorn people bought. The results are in the table below.

a) Draw and label a pie chart to represent the information.

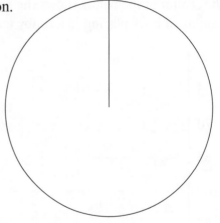

Type of popcorn	Number sold
Plain	12
Salted	18
Sugared	9
Toffee	21

[4]

Another survey was carried out to find out which flavour of ice cream people bought. The results are shown in the pie chart below.

Chris compares the two pie charts and says,

"The results show that more people chose strawberry ice cream than toffee popcorn."

b) Explain whether or not Chris is right.

....

....

....

....

[1]

[Total 5 marks]

Score: ☐

8

Section Seven — Probability and Statistics

Scatter Graphs

1   15 pupils in a class study both Spanish and Italian.
Their end of year exam results are shown on the scatter graph below.

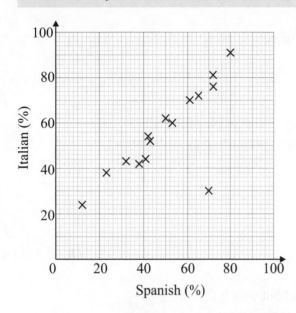

Italian (%)

Spanish (%)

a) Circle the point that doesn't follow the trend.

[1]

b) Describe the strength and type of correlation shown by the points that do follow the trend.

....

....

[1]

c) Draw a line of best fit for the data.

[1]

[Total 3 marks]

2   The scatter graph below shows the heights and weights of boys playing in a rugby team.

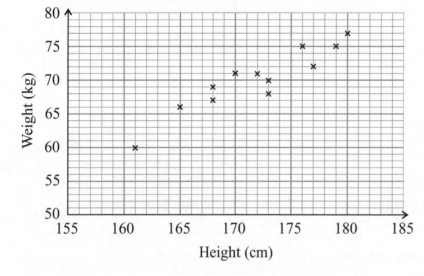

Weight (kg)

Height (cm)

Two more boys join the team. Their heights and weights are shown in this table.

Player	Height (cm)	Weight (kg)
13	169	70
14	183	76

a) Add the information in the table to the scatter graph.

[1]

b) What fraction of the players have a height of less than 170 cm?

....

[1]

c) Describe the relationship between the height and weight of the players.

....

....

[1]

[Total 3 marks]

Section Seven — Probability and Statistics

3 A furniture company is looking at how effective their advertising is.
They are comparing how much they spent on advertising in random months with
their total sales value for that month. This information is shown on the graph below.

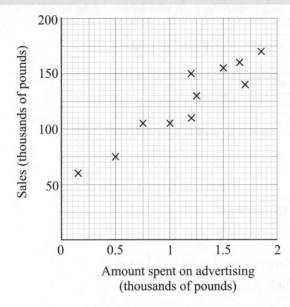

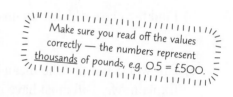

Make sure you read off the values
correctly — the numbers represent
thousands of pounds, e.g. 0.5 = £500.

a) Use a line of best fit to estimate how much the company would be likely to spend
on advertising in a month where they sold £125 000 worth of furniture.

£ ....
[2]

b) Estimate the monthly sales value if the company spends £600 on advertising.

£ ....
[1]

c) Comment on the reliability of your estimates in parts a) and b).

....
[1]

d) Next month the company plans to spend £3000 on advertising.
They will use the trend in the data above to predict the sales value for next month.

Explain why this prediction might not be reliable.

....

....
[1]

e) An employee says that 'increasing the amount spent on advertising causes sales to increase'.

Explain whether the graph above proves this statement.

....

....
[2]

[Total 7 marks]

Score:

13

Grouped Frequency Tables

1   For a science experiment, Bill planted 10 seeds and measured their growth to the nearest cm after 12 days. His results are shown in the table below.

Growth in cm	Number of plants
$0 \leq x \leq 2$	2
$3 \leq x \leq 5$	4
$6 \leq x \leq 8$	3
$9 \leq x \leq 11$	1

a) Find the modal class.

....
[1]

b) Find the class which contains the median.

....
[1]

c) Bill works out that the mean height after 12 days is 12 cm.
Explain why Bill must have made a mistake.

....

[1]
[Total 3 marks]

2   The table shows the times it took 32 pupils at a school to run a 200 m sprint.

a) Calculate an estimate for the mean time taken.

Time (t seconds)	Frequency	Mid-interval value	Frequency × Mid-interval value
$22 < t \leq 26$	4	$(22 + 26) \div 2 = 24$	$4 \times 24 = $ ....
$26 < t \leq 30$	8		
$30 < t \leq 34$	13		
$34 < t \leq 38$	6		
$38 < t \leq 42$	1		
Total			

Estimate of mean = ... ÷ .... = ...

.... seconds
[4]

All pupils with a time of 34 seconds or less qualified for the next round.

b) Anya says that fewer than 20% of the pupils failed to qualify for the next round.
Comment on Anya's statement and show working to support your answer.

....

....

[2]
[Total 6 marks]

Score:

9

Candidate Surname		Candidate Forename(s)	
Centre Number	Candidate Number	Candidate Signature	

GCSE

Mathematics
Paper 1 (Non-Calculator)

Foundation Tier

Practice Paper
Time allowed: 1 hour 30 minutes

You must have:
Pen, pencil, eraser, ruler, protractor, pair of compasses.
You may use tracing paper.

You are not allowed to use a calculator.

Instructions to candidates
- Use **black** ink to write your answers.
- Write your name and other details in the spaces provided above.
- Answer **all** questions in the spaces provided.
- In calculations show clearly how you worked out your answers.
- Do all rough work on the paper.

Information for candidates
- The marks available are given in brackets at the end of each question.
- You may get marks for method, even if your answer is incorrect.
- There are 26 questions in this paper. There are no blank pages.
- There are 80 marks available for this paper.

Get the answers online
Worked solutions to this practice paper are available online for you to download or print.
Go to **www.cgpbooks.co.uk/gcsemathsanswers** to get them.

Answer ALL the questions.

Write your answers in the spaces provided.

You must show all of your working.

1   Write 0.113 as a fraction.
    Circle your answer.

$$\frac{113}{100} \qquad\qquad \frac{113}{10\,000} \qquad\qquad \frac{113}{1000} \qquad\qquad \frac{13}{100}$$

[Total 1 mark]

2   Write the ratio 40 : 25 in its simplest form.

....

[Total 1 mark]

3   Eight points are shown plotted on the grid.

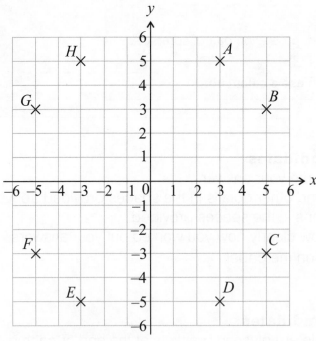

(a)   Circle the point that has coordinates (−5, −3).

    C   E   F   G

[1]

(b)   Circle the equation of the straight line that passes through points *A* and *D*.

    x = 3   *x* + *y* = 3   *y* = 3*x*   *y* = 3

[1]

[Total 2 marks]

1

4 Convert 3.97 km into m.

.... m

[Total 1 mark]

5 Karl has five number cards.

$\boxed{-6}$ $\boxed{6}$ $\boxed{-8}$ $\boxed{-12}$ $\boxed{2}$

(a) Write Karl's number cards in order, starting with the lowest.

lowest ...., ..., ..., ..., ... highest

[1]

(b) Use two of Karl's number cards to make this calculation correct.

.... $-$ .... $= 10$

[1]

[Total 2 marks]

6 Beth has some 5p and 10p coins.

Coin	Number
5p	28
10p	41

She changes her coins for 50p coins at the bank.

How many 50p coins does she receive?

....

[Total 3 marks]

7 Calculate

$$\frac{1.2 - 0.2 \times 4}{0.05}$$

....

[Total 2 marks]

8

> If you add a multiple of 3 to a multiple of 6, you always get a multiple of 9.

Give an example to show that this statement is not true.

....

....

[Total 1 mark]

---

9 (a) Simplify $11a + 5b - 2a + 2b$

Circle your answer.

$13a + 3b$ $\qquad$ $9a + 7b$ $\qquad$ $13a + 7b$ $\qquad$ $16ab$

[1]

(b) Simplify $2a \times 3a$

Circle your answer.

$5a$ $\qquad$ $6a$ $\qquad$ $5a^2$ $\qquad$ $6a^2$

[1]

[Total 2 marks]

---

10 Chloe invests £300 in a bank account.
The account pays 2% simple interest each year.

Work out how much money she has in her account after 4 years.

£ ....

[Total 3 marks]

11 The dual bar chart below shows the favourite sports of the pupils in a class.

One bar is missing.

There are 30 children in the class.

(a) Draw the missing bar to show the number of boys whose favourite sport is Hockey.

[2]

(b) One child is chosen at random from the class.
Find the probability that their favourite sport is swimming.

....

[2]

(c) What is the ratio of the number of boys who chose swimming to the number of girls who chose tennis? Give your answer in its simplest form.

....

[2]

[Total 6 marks]

12 Decide whether the sequence is an arithmetic or geometric progression, and write down in words the rule to get from one term to the next.

2, 8, 32, 128...

Arithmetic ☐   Geometric ☐

Rule: ....

....

[Total 2 marks]

4

13 The diagram shows the first four patterns in a sequence.

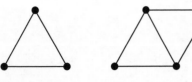 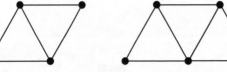 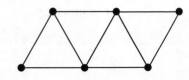

  Pattern 1   Pattern 2   Pattern 3   Pattern 4

(a) Complete the table.

	Number of triangles	Number of dots	Number of lines
Pattern 1	1	3	3
Pattern 2	2	....	5
Pattern 3	....	5	....
Pattern 4	4	....	....

[1]

(b) Work out the number of lines in pattern 10.

....

[2]

(c) (i) Find a formula for the number of dots D in Pattern n.

....

[2]

  (ii) Find the number of dots in pattern 200.

....

[1]

[Total 6 marks]

14 A theatre sells three types of tickets.

Ticket type	Cost
Adult	£9
Child	£5
Senior	£6.50

The pictogram shows the number of tickets of each type sold for one performance.

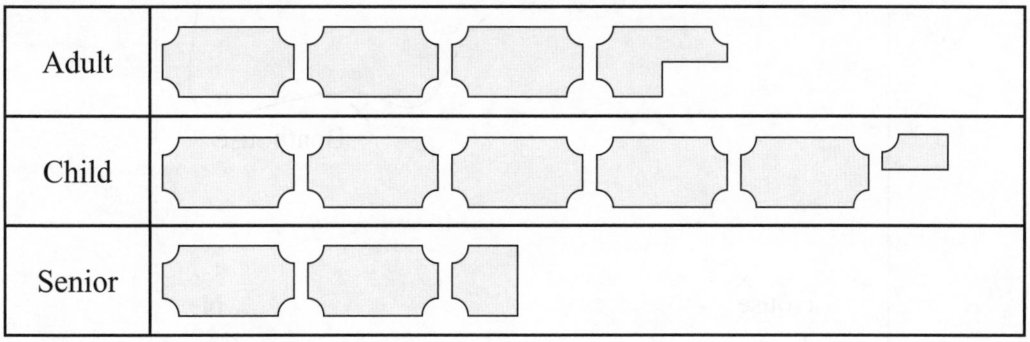

Adult	
Child	
Senior	

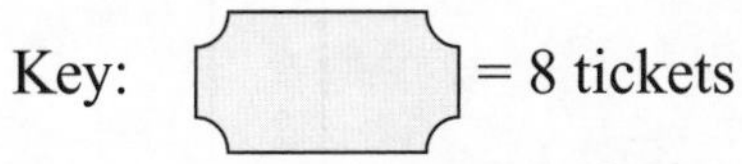

Key: = 8 tickets

How much money did the theatre make from all the ticket sales for this performance?

£ ....

[Total 6 marks]

15 The scale drawing shows the gardens of a country house.

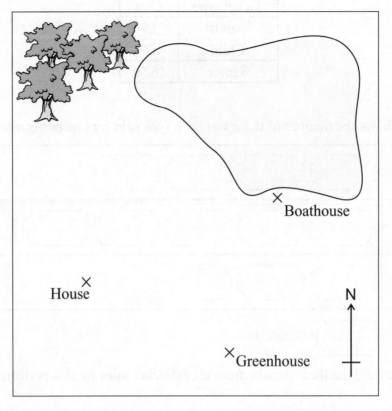

Scale: 1 cm = 100 metres

(a) Find the three-figure bearing of the boathouse from the house.

.................°

[1]

(b) Find the actual distance from the boathouse to the greenhouse.

.... metres

[2]

(c) A summerhouse is

450 metres from the house
700 metres from the greenhouse

Plot with a cross (×) the position of the summerhouse on the map.
Do not rub out your construction lines.

[2]

[Total 5 marks]

16 Angie makes wedding cakes with three tiers.

She needs 800 grams of sultanas to make the bottom tier of a cake.
The middle tier needs 75% of the ingredients required for the bottom tier.
The top tier needs 50% of the ingredients of the bottom tier.

Angie needs to make five wedding cakes. She has 8 kilograms of sultanas.

Does Angie have enough sultanas to make five cakes? Show your working.

[Total 5 marks]

17 Carrots cost 69p per kilogram.
Ahmed buys 2.785 kilograms of carrots.

(a) Estimate the cost, in pence, of his carrots.
Show the numbers you use to work out your estimate.

.... p

[2]

(b) Is your estimate in (a) bigger or smaller than the actual cost?
Tick the correct answer.

Bigger ☐   Smaller ☐

Explain your answer.

....

....

[1]

[Total 3 marks]

8

18 The graph can be used to convert between pounds (£) and Australian dollars ($).

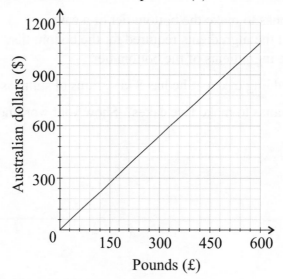

Jack goes on holiday to Australia and China.

(a) He changes £300 into Australian dollars ($).
How many Australian dollars does he get for £300?

$ ....
[1]

(b) He spends $390 of his money while in Australia.
He converts the rest of his Australian dollars into Chinese yuan using the exchange rate

1 Australian dollar = 6 Chinese yuan

How many Chinese yuan does Jack get?

.... yuan
[2]
[Total 3 marks]

19 Work out the value of k if

$$k \times 3^{-2} = 4$$

$k =$ ....
[Total 2 marks]

9

20 The diagram shows an isosceles triangle.

Three of these isosceles triangles fit together with three squares around a point O.

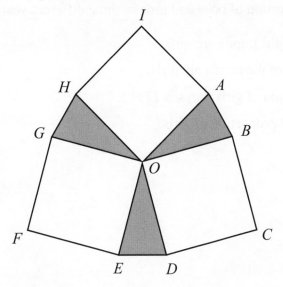

Show clearly that angle $OAB = 75°$.

[Total 4 marks]

21 Work out $1\frac{2}{3} \times 1\frac{5}{8}$. Give your answer as a mixed number.

....

[Total 3 marks]

22 Write 594 000 000 000 in standard form.

....

[Total 1 mark]

23 A school records the proportion of boys and girls in three different year groups.

In Year 9, $\frac{9}{20}$ of the pupils are girls.

In Year 10, 49% of the pupils are girls.

In Year 11, the ratio of girls : boys is 12 : 13.

Which year has the largest proportion of girls?

....

[Total 3 marks]

24 The diagram shows a circle *A* and a sector *B*.

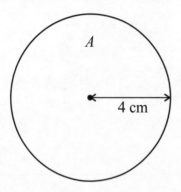
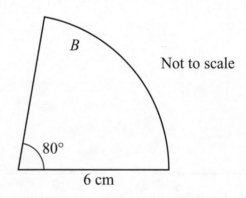

Not to scale

Show that the area of *A* is twice the area of *B*.

....

[Total 4 marks]

25 Quadrilateral *ABCD* is made up of two right-angled triangles, *ABC* and *ACD*.

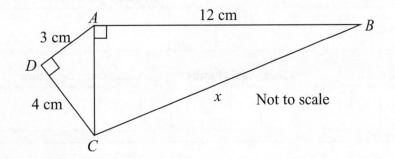

(a) Find the length of the side labelled *x*.

.... cm

[4]

(b) Find the area of quadrilateral *ABCD*.

.... cm^2

[2]

[Total 6 marks]

26 Solve the simultaneous equations

$$3x + 2y = 17$$
$$2x + y = 10$$

x = ...

y = ...

[Total 3 marks]

[TOTAL FOR PAPER = 80 MARKS]

12

Practice Paper 1

Candidate Surname	Candidate Forename(s)

Centre Number	Candidate Number	Candidate Signature

GCSE

Mathematics
Paper 2 (Calculator)

Foundation Tier

Practice Paper
Time allowed: 1 hour 30 minutes

You must have:
Pen, pencil, eraser, ruler, protractor, pair of compasses.
You may use tracing paper.

You **may use** a calculator.

Instructions to candidates
- Use **black** ink to write your answers.
- Write your name and other details in the spaces provided above.
- Answer **all** questions in the spaces provided.
- In calculations show clearly how you worked out your answers.
- Do all rough work on the paper.
- Unless a question tells you otherwise, take the value of π to be 3.142, or use the π button on your calculator.

Information for candidates
- The marks available are given in brackets at the end of each question.
- You may get marks for method, even if your answer is incorrect.
- There are 28 questions in this paper. There are no blank pages.
- There are 80 marks available for this paper.

Get the answers online

Worked solutions to this practice paper are available online for you to download or print.
Go to **www.cgpbooks.co.uk/gcsemathsanswers** to get them.

Answer ALL the questions.

Write your answers in the spaces provided.

You must show all of your working.

---

1   Write $\frac{3}{5}$ as a percentage.

Circle your answer.

6%   30%   15%   60%

---

2   A function is represented by this number machine.

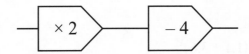

The **output** of the machine is 20. Circle the input.

8   12   14   36

[Total 1 mark]

---

3   Complete this bill.

Barbara's Café			
Menu Item	**Number Ordered**	**Cost per Item**	**Total**
Tea	2	£1.25	£2.50
Coffee	....	£1.60	£9.60
Cake	4	£	£5.20
Tip			£2.50
		Total cost	£ ....

[Total 3 marks]

1

4 (a) Draw a line to match each shape to its number of **surfaces**.

cone   4

                          3

sphere

                          2

cylinder   1

[2]

(b) Write down the number of **vertices** for a triangular prism.

....

[1]

[Total 3 marks]

---

5 (a) Calculate

$$\sqrt{12.2} + (1.1 + 3.6)^3$$

Write down all the digits on your calculator.

....

[1]

(b) Round your answer to (a) to two decimal places.

....

[1]

[Total 2 marks]

---

6 Kamil has these four number cards.

3   5   4   6

List the eight even numbers greater than 4000 Kamil can make by rearranging all four cards.

[Total 2 marks]

7   Circle the number below that is both a square number and a cube number.

    16   27   64   25   8   100

8   (a)   Expand

$$4(a + 2)$$

....
[1]

    (b)   Factorise

$$y^2 + 5y$$

....
[1]
[Total 2 marks]

9   The ages (in years) of seven children are

    6   12   9   6   5   7   11

    (a)   Find the median age.

....
[1]

    (b)   Find the mean age.

....
[2]
[Total 3 marks]

10 An equilateral triangle *T* is shown on the grid.

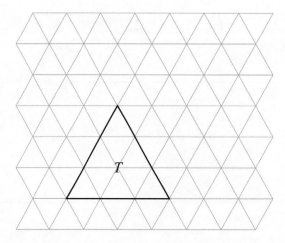

(a) Another triangle congruent to *T* is joined to *T* to form a quadrilateral.
Write down the number of lines of symmetry of the quadrilateral.

....

[1]

(b) Show on the grid how four triangles congruent to *T* can be joined together
to form a shape with rotational symmetry of order 3.

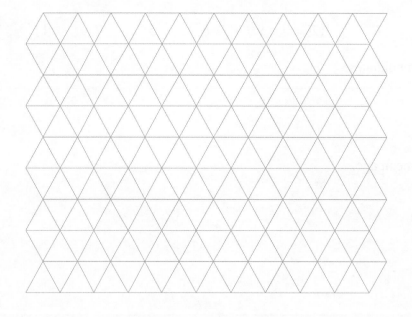

[1]

[Total 2 marks]

11 Work out 185% of £3500.

£ ....

[Total 2 marks]

12 Sandra attends a job interview at a school.

The school refunds her travelling expenses if she uses the cheapest possible method of transport.

There are two methods of transport that Sandra can use to attend the interview.

Method 1: By car
Sandra lives 27 miles from the school.

Method 2: By car and train
Sandra lives 4 miles from the station.
The cost of a return train ticket is £17.60.

The school refunds car travel at a rate of 40p per mile.

Which method should Sandra use to travel to her interview and home again if she wants a refund for her expenses?

Show how you work out your answer.

[Total 3 marks]

13 A car park contains 28 cars and 16 motorbikes.

$\frac{3}{4}$ of the cars and $\frac{3}{8}$ of the motorbikes are red.

A red vehicle is chosen at random.

What is the probability that it is a car? Give your answer as a fraction in its simplest form.

....

[Total 3 marks]

14 (a) Complete the table of values for $y = 7 - 2x$.

x	-2	-1	0	1	2	3	4
y		9	7			1	

[2]

(b) Draw the graph of $y = 7 - 2x$ for values of x between -2 and 4.

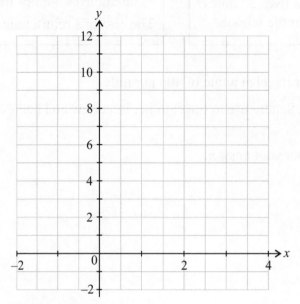

[2]

(c) What is the gradient of the line $y = 7 - 2x$?
Circle your answer.

-2 -1 2 7

[1]

[Total 5 marks]

---

15 Solve the equation $3(2x - 4) = 2x + 8$

$x = $ ....

[Total 3 marks]

---

16 Kieron works out $5 \times \dfrac{2}{3}$ and gets the answer $\dfrac{10}{15}$.

Explain what mistake Kieron has made in calculating his answer.

....

....

[Total 1 mark]

17 The budget airline 'Fly By Us' produces this graph to show how their passenger numbers have increased.

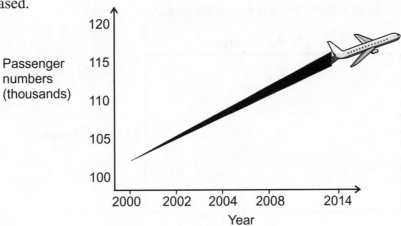

Give three different criticisms of the graph.

Criticism 1 ....

....

Criticism 2 ....

....

Criticism 3 ....

....

[Total 3 marks]

18 *AC* and *DG* are straight lines.

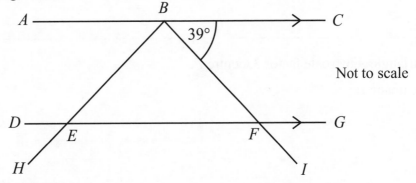

Not to scale

BH is perpendicular to *BI*.

Work out the size of angle *DEH*.
Show how you work out your answer.

.......................°

[Total 3 marks]

Practice Paper 2

19 Jimmy has a rectangular vegetable garden measuring $(3x + 1)$ metres by $(2x - 3)$ metres.

Jimmy wants to put a fence around the outside of the garden.

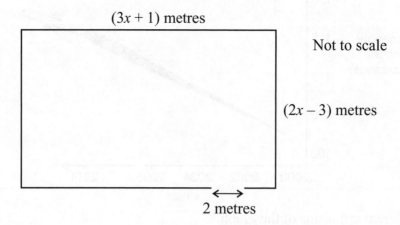

(3x + 1) metres

Not to scale

(2x – 3) metres

2 metres

He needs a 2 metre gap along one edge so that he can get in and out.

(a) Show that the length of the fence, L m, is given by the formula $L = 10x - 6$.

[2]

(b) Show that L is always an even number when x is a whole number.

[2]

[Total 4 marks]

20 Enlarge triangle A by scale factor 3, centre P.

Label the image B.

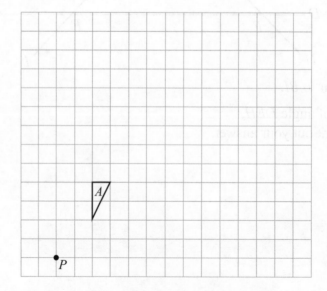

[Total 2 marks]

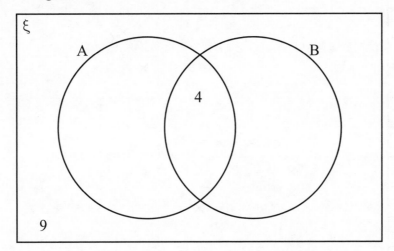

21 $\xi = \{1, 2, 3, \ldots, 10\}$
$A = \{x : 2 < x \le 6\}$
$B = \{x : x \text{ is a factor of } 12\}$

Complete the Venn diagram to show the elements of each set.

[Total 3 marks]

22 Orange juice and lemonade are mixed in the ratio 3 : 5 to make orangeade.

Orange juice costs £1.60 per litre.
Lemonade costs £1.20 per litre.

What is the cost of making 18 litres of orangeade?

£ ....

[Total 4 marks]

23 Make x the subject of the formula

$$y = \frac{x^2 - 2}{5}$$

....

[Total 2 marks]

24 The scatter graph shows the maximum power (in kW)
and the maximum speed (in km/h) of a sample of cars.

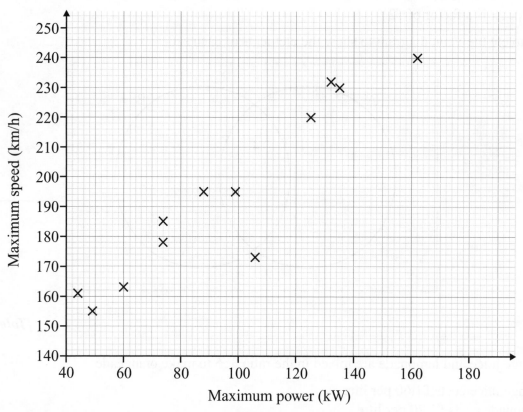

(a) One of the cars has a maximum speed of 220 km/h.
Write down the maximum power of this car.

.... kW

[1]

(b) One of the points is an outlier as it does not fit in with the trend.
Draw a ring around this point on the graph.

[1]

(c) Ignoring the outlier, describe the correlation shown on the scatter graph.

.... correlation

[1]

(d) A different car has a maximum power of 104 kW.
By drawing a suitable line on your scatter graph, estimate the maximum speed of this car.

.... km/h

[2]

(e) Explain why it may not be reliable to use the scatter graph to estimate
the maximum speed of a car with a maximum power of 190 kW.

....

....

[1]

[Total 6 marks]

10

Practice Paper 2

25 Ollie and Amie each have an expression.

| Ollie |
| $(x + 4)^2 - 1$ |

| Amie |
| $(x + 5)(x + 3)$ |

Show clearly that Ollie's expression is equivalent to Amie's expression.

[Total 3 marks]

26 A company consists of 80 office assistants and a number of managers.

The pie chart shows how the 80 office assistants travel to work.

Office Assistants

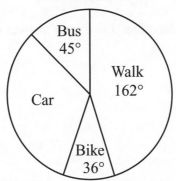

(a) How many office assistants travel to work by car?

....
[2]

18 of the managers travel by car.
Overall, 40% of the people in the company travel by car.

(b) Work out how many people there are in the company.

....
[2]

[Total 4 marks]

27 The diagram shows a solid aluminium cylinder and a solid silver cube.

Cylinder (aluminium)   Cube (silver)

 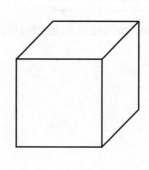

Not to scale

- The volume of the cylinder is 1180 cm³.
- The cylinder and the cube have the same mass.
- The density of aluminium is 2.7 g/cm³ and the density of silver is 10.5 g/cm³.

(a) Calculate the mass of the cylinder.

.... g

[2]

(b) Calculate the side length of the cube. Give your answer correct to two significant figures.

.... cm

[4]

[Total 6 marks]

28 The diagram shows a right-angled triangle.

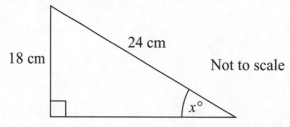

18 cm   24 cm   Not to scale

$x°$

Calculate the value of x. Give your answer correct to 1 decimal place.

$x =$ ....

[Total 3 marks]

[TOTAL FOR PAPER = 80 MARKS]

12

Practice Paper 2

Candidate Surname

Candidate Forename(s)

Centre Number

Candidate Number

Candidate Signature

GCSE

Mathematics

Foundation Tier

Paper 3 (Calculator)

Practice Paper
Time allowed: 1 hour 30 minutes

You must have:
Pen, pencil, eraser, ruler, protractor, pair of compasses.
You may use tracing paper.

You may use a calculator.

Instructions to candidates

- Use **black** ink to write your answers.
- Write your name and other details in the spaces provided above.
- Answer **all** questions in the spaces provided.
- In calculations show clearly how you worked out your answers.
- Do all rough work on the paper.
- Unless a question tells you otherwise, take the value of π to be 3.142,
  or use the π button on your calculator.

Information for candidates

- The marks available are given in brackets at the end of each question.
- You may get marks for method, even if your answer is incorrect.
- There are 28 questions in this paper. There are no blank pages.
- There are 80 marks available for this paper.

Get the answers online
Worked solutions to this practice paper are available online for you to download or print.
Go to **www.cgpbooks.co.uk/gcsemathsanswers** to get them.

Answer ALL the questions.

Write your answers in the spaces provided.

You must show all of your working.

1   Write one of the signs <, =, or > on each answer line to make a true statement.

$$0.4 \quad \dots\dots\dots \quad 0.34$$

$$\frac{3}{4} \quad \dots\dots\dots \quad 0.75$$

$$7\% \quad \dots\dots\dots \quad 0.7$$

[Total 2 marks]

2   The diagram shows part of a number line.

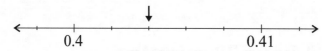

Circle the number the arrow points to.

0.42   0.44   0.402   0.404

[Total 1 mark]

3   Round 20 758 to the nearest 100.

....
[Total 1 mark]

4   What number is 12 less than –4.2?

....
[Total 1 mark]

5   Circle the number below that has exactly four factors.

2   3   5   8   9   12

[Total 1 mark]

1

6 24 pupils take a violin exam. Their marks are shown below.

110	103	115	134	121	98
128	112	107	112	125	132
114	102	125	93	120	120
106	111	99	98	127	115

The certificate each pupil receives depends upon their mark.

Result of exam	Mark
Fail	Under 100
Pass	100 – 119
Merit	120 – 129
Distinction	130 and above

(a) Complete the table to show the number of pupils achieving each result.
The first row has been filled in for you.

Result of exam	Tally	Frequency
Fail	||||	4
Pass		
Merit		
Distinction		
	Total:	**24**

[2]

(b) What fraction of the pupils failed the exam? Give your fraction in its simplest form.

....

[2]

(c) Draw on the grid a suitable diagram to show the number of pupils achieving each result.

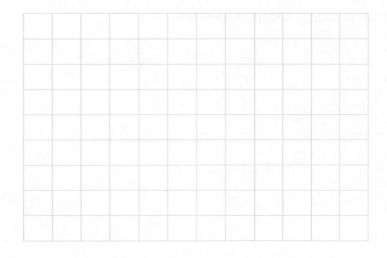

[3]

[Total 7 marks]

2

7   A pencil case contains 10 coloured pencils.

1 pencil is yellow.
2 pencils are red.
The other pencils are either green or blue.

Carla picks one coloured pencil at random.
She has the same chance of picking a green pencil as a red pencil.

Circle the word that describes the probability of picking:

(a)   a black pencil,

impossible   unlikely   evens   likely

[1]

(b)   a blue pencil.

impossible   unlikely   evens   likely

[1]

[Total 2 marks]

8   Here are the names of four types of quadrilateral.

Parallelogram   Square   Trapezium   Kite

Choose from this list the quadrilateral that has:

(a)   exactly one pair of parallel sides,

....

[1]

(b)   no lines of symmetry, but rotational symmetry of order 2.

....

[1]

[Total 2 marks]

9   Circle the vector that translates a shape 5 units **left**.

$$\begin{pmatrix} -5 \\ 0 \end{pmatrix} \qquad \begin{pmatrix} 5 \\ 0 \end{pmatrix} \qquad \begin{pmatrix} 0 \\ 5 \end{pmatrix} \qquad \begin{pmatrix} 0 \\ -5 \end{pmatrix}$$

[Total 1 mark]

10 Rick multiplies three different numbers together and gets 90.
One of his numbers is a square number, and the other two are prime numbers.
What are the three numbers he uses?

....

[Total 3 marks]

---

11 Nigel sees this recipe for cupcakes.

Recipe for 12 cupcakes
140 grams butter
140 grams flour
132 grams sugar
2 eggs
1 tablespoon milk

Nigel wants to make 30 cupcakes. How much sugar does he need?

.... g

[Total 2 marks]

---

12 Ajay buys some packets of ginger biscuits.
Jane buys some packets of shortbread biscuits.

Ginger biscuits
contains 12 ginger biscuits

Shortbread biscuits
contains 10 shortbread biscuits

Ajay and Jane buy the same number of biscuits.

What is the smallest number of packets of shortbread biscuits Jane could have bought?

.... packets

[Total 3 marks]

13 Mary is preparing cream teas for 30 people.

Each person needs 2 scones, 1 tub of clotted cream and 1 small pot of jam.

She has £35 to buy everything.

  A pack of 10 scones costs £1.35
  A pack of 6 tubs of clotted cream costs £2.95
  Each small pot of jam costs 40p

Will she have enough money? Show how you work out your answer.

....

[Total 5 marks]

14 The grid shows part of two shapes, *A* and *B*.

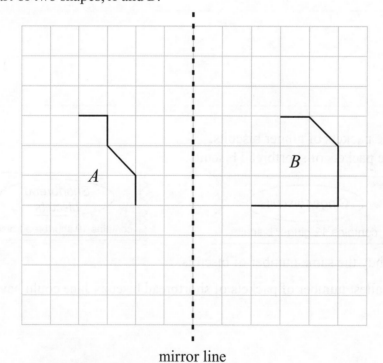

mirror line

B is the reflection of *A* in the mirror line.

Complete both shapes.

[Total 2 marks]

5

Practice Paper 3

15 The diagram shows an object made from 8 centimetre cubes.

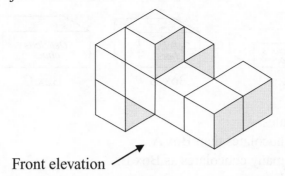

Front elevation

Draw on the grids below the plan view and the front elevation of the object.

Plan view   Front elevation

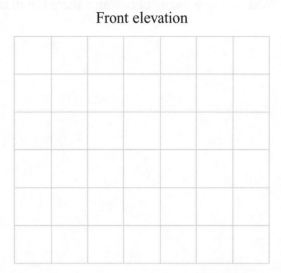

[Total 2 marks]

16 Two congruent trapeziums and two triangles fit inside a square of side 12 cm as shown.

12 cm

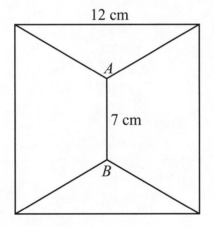

Not to scale

A

7 cm

B

$AB = 7$ cm

Work out the area of each trapezium.

.... cm²

[Total 2 marks]

Practice Paper 3

138

17 A chocolate manufacturer makes boxes of chocolates in three different sizes.

Box A   Box B   Box C

Box A contains c chocolates.
Box B contains 4 more chocolates than Box A.
Box C contains twice as many chocolates as Box B.
Altogether there are 60 chocolates.

Work out how many chocolates there are in each box.

Box A: ....

Box B: ....

Box C: ....

[Total 5 marks]

7

Practice Paper 3

18 Simplify

(a) $y \times y \times y$

....
[1]

(b) $n^6 \div n^2$

....
[1]

(c) $(a^4)^3$

....
[1]

[Total 3 marks]

---

19 The diagram shows a ramp placed against two steps.

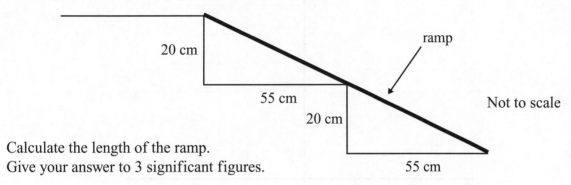

20 cm

55 cm

20 cm

55 cm

ramp

Not to scale

Calculate the length of the ramp.
Give your answer to 3 significant figures.

.... cm
[Total 3 marks]

---

20 A route between Guilford and Bath has a distance of 180 kilometres.
Dave drives from Guilford to Bath. He takes 3 hours.

Olivia drives the same route. Her average speed is 15 kilometres per hour faster than Dave's.

(a) How long does it take Olivia to drive from Guilford to Bath?
Give your answer in hours and minutes

.... hours .... minutes
[3]

(b) Why is it important to your calculation that Olivia drives the same route as Dave?

....

....
[1]

[Total 4 marks]

21 (a) Complete the table of values for $y = x^2 + x - 2$.

x	-3	-2	-1	0	1	2	3
y		0	-2	-2			10

[2]

(b) Draw on the grid the graph of $y = x^2 + x - 2$ for values of x between -3 and 3.

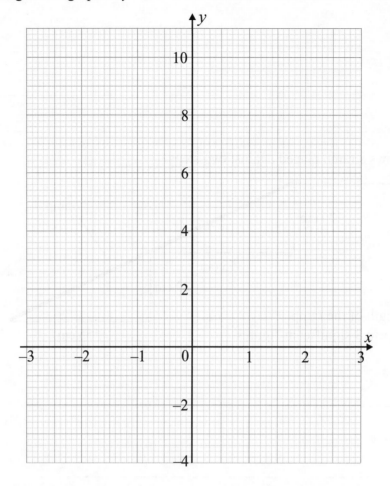

[2]

[Total 4 marks]

22 The ratio of angles in a triangle is $2:3:5$.
Show that it is a right-angled triangle.

[Total 3 marks]

Practice Paper 3

23 The values of four houses at the start of 2013 are shown.

House 1	House 2	House 3	House 4
£120 000	£144 000	£145 000	£150 000

(a) Which house has a value 25% higher than House 1?

House ....

[1]

(b) At the start of 2015, the value of House 2 is £161 280.
Find the percentage increase in the value of House 2.

.... %

[3]

[Total 4 marks]

24 Anna and Carl each think of a sequence of numbers.

Anna's sequence

4th term = 17

Term-to-term rule is
Add 3

Carl's sequence

Term-to-term rule is
Add 6

The 1st term of Anna's sequence is double the 1st term of Carl's sequence.

Work out the 5th term of Carl's sequence.

....

[Total 3 marks]

25 (a) Factorise $x^2 + 7x - 18$.

....

[2]

(b) Solve the equation $x^2 + 7x - 18 = 0$.

$x =$ .... or $x =$ ....

[1]

[Total 3 marks]

26 George has two fair spinners.

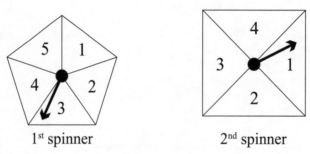

1st spinner   2nd spinner

He spins each spinner once and records whether the score is an odd or an even number.

(a) Complete the tree diagram to show the probabilities.

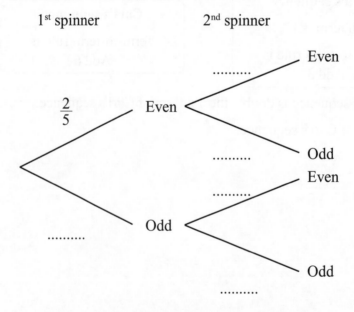

(b) Work out the probability that George spins two odd numbers.

....

[2]

[Total 4 marks]

11

27 The line L passes through the points $(-2, -7)$ and $(3, 8)$.
Find the equation of line L.

....

[Total 4 marks]

28 The grouped frequency table below shows the weights of 25 rabbits in a pet shop.

Weight (w g)	Frequency
$800 \leq w < 1000$	5
$1000 \leq w < 1200$	8
$1200 \leq w < 1400$	9
$1400 \leq w < 1600$	3

Estimate the mean weight.

.... g

[Total 3 marks]

[TOTAL FOR PAPER = 80 MARKS]

Section One — Number

Page 3: Types of Number and BODMAS

1. a) $11 + 14 \div 2 = 11 + 7 = 18$
   [2 marks available — 1 mark for doing the calculation steps in the correct order, 1 mark for the correct answer]
   b) $(20 - 15) \times (4 + 6) = 5 \times 10 = 50$
   [2 marks available — 1 mark for doing the calculation steps in the correct order, 1 mark for the correct answer]

2. a) 81 *[1 mark]*
   b) 64 *[1 mark]*

3. $18, \frac{12}{6} (= 2), -22$
   [2 marks available — 2 marks for all three integers circled, or 1 mark for two integers circled. Lose 1 mark for each non-integer circled (up to a maximum of 2).]

4. $\dfrac{197.8}{\sqrt{0.01 + 0.23}} = \dfrac{197.8}{\sqrt{0.24}} = \dfrac{197.8}{0.489897948...}$
   $= 403.7575593$
   [2 marks available — 1 mark for correct working, 1 mark for correct answer]

Pages 4-5: Wordy Real-Life Problems

1. $522 - (197 + 24) = 301$
   [2 marks available — 1 mark for subtracting the two numbers from 522, 1 mark for the correct answer]

2. $£15 - £8.50 + £20 - £18 = £8.50$, so he would not have £10 to give to his sister.
   [2 marks available — 1 mark for adding and subtracting the correct amounts to work out how much he would have left, 1 mark for a correct conclusion based on the correct amount]

3. Paying separately: Parvati: £2.30 + £1.40 = £3.70
   Zayn: £1.90 + £1.80 = £3.70
   Total: £3.70 + £3.70 = £7.40
   Paying together: £3.40 + £1.90 + £1.40 = £6.70
   They would save £7.40 – £6.70 = £0.70 if they paid together.
   [3 marks available — 1 mark for finding the total cost when paying separately, 1 mark for finding the total cost when paying together, 1 mark for finding the difference]
   For this one, you have to spot that they could use the breakfast deal if they combined their orders.

4. Total cost = £2.15 + £2.40 + £2.40 = £6.95
   Change = £10 – £6.95 = £3.05
   [2 marks available — 1 mark for adding amounts and subtracting from £10, 1 mark for the correct answer]

5. Total amount of drink = 500 ml + 216 ml = 716 ml
   716 ml ÷ 2 = 358 ml each *[1 mark]*
   500 ml – 358 ml = 142 ml, so Theo gives 142 ml of his drink to Poppy *[1 mark]*.
   [2 marks available in total — as above]

6. Total miles travelled = $(30 \times 2) + (28 \times 2) + (39 \times 2) + (40 \times 2)$
   $= 60 + 56 + 78 + 80$
   $= 274$ miles
   Expenses for miles travelled = 274 × 30p = 8220p = £82.20
   Expenses for food = 4 × £8 = £32
   Total expenses = £82.20 + £32 = £114.20
   [4 marks available — 1 mark for finding total miles, 1 mark for multiplying total miles by 30 or 0.3(0), 1 mark for finding food expenses, 1 mark for the correct final answer]

Pages 6-7: Multiplying and Dividing

1. $12 \times 15 = 3 \times (4 \times 15) = 3 \times 60$ *[1 mark]* = 180 *[1 mark]*
   [2 marks available in total – as above]

2. 468 × 38 = 17 784p = £177.84 *[1 mark]*
   402 × 44 = 17 688p = £176.88 *[1 mark]*
   £177.84 – £176.88 = £0.96 *[1 mark]*
   [3 marks available in total — as above]

3. E.g. 672 ÷ 2 = 336
   $336 \div 6 = 6\overline{)3^3 3^3 6}$ $\begin{smallmatrix}0\,5\,6\end{smallmatrix}$ = 56
   56 ÷ 7 = 8
   So the fourth number is 8
   [3 marks available — 1 mark for a correct method, 1 mark for at least one correct calculation, 1 mark for the correct answer]

4. £200 – £5 = £195
   $15\overline{)1^1 9^4 5}$ $\begin{smallmatrix}0\,1\,3\end{smallmatrix}$, so each ticket costs £13
   [3 marks available — 1 mark for subtracting £5 from £200, 1 mark for dividing £195 by 15, 1 mark for the correct final answer]

5. a)
   $$\begin{array}{r} 29 \\ \times\ 19 \\ \hline 261 \\ +\ 290 \\ \hline 551 \end{array}$$
   29 × 1.9 has one digit after the decimal point, so 29 × 1.9 = 55.1 = £55.10
   [2 marks available — 1 mark for a correct method, 1 mark for the correct answer]
   b) 29 + 57 = 86
   $6\overline{)8^2 6}$ $\begin{smallmatrix}1\,4\ r\ 2\end{smallmatrix}$
   So there will be at least 15 groups.
   [2 marks available — 1 mark for finding the total number of students, 1 mark for dividing to find the correct answer]

6. a) 1 worker eats 2 biscuits per day
   20 workers will eat 20 × 2 = 40 biscuits per day *[1 mark]*
   So they will eat 40 × 7 = 280 biscuits per week *[1 mark]*
   [2 marks available in total — as above]
   b) They will need 280 ÷ 14 = 20 packets per week *[1 mark]*
   20 packets will cost 20 × £1.30 = £26 per week *[1 mark]*
   [2 marks available in total — as above]

7. a) $14 \div 0.7 = \dfrac{14}{0.7} = \dfrac{140}{7}$
   $7\overline{)1^1 4\,0}$ $\begin{smallmatrix}0\,2\,0\end{smallmatrix}$ = 20
   [2 marks available — 1 mark for a correct method, 1 mark for the correct answer]
   b) $2.76 \div 0.12 = \dfrac{2.76}{0.12} = \dfrac{276}{12}$
   $12\overline{)2^2 7^3 6}$ $\begin{smallmatrix}0\,2\,3\end{smallmatrix}$ = 23
   [2 marks available — 1 mark for a correct method, 1 mark for the correct answer]

8. Slices of pizza he needs = 15 × 3 = 45 *[1 mark]*
   Number of pizzas he needs = 45 ÷ 8 *[1 mark]*
   = 5 r 5
   So he needs 6 pizzas *[1 mark]*
   A 300 g packet of crisps is enough for 300 ÷ 25 = 12 people *[1 mark]*. So James needs 2 packets of crisps *[1 mark]*
   [5 marks available in total — as above]
   Another way of working out the number of packets of crisps needed is to find the total amount needed in grams and then divide by 300 g.

Page 8: Negative Numbers

1. –9, –8, –3, 0, 3, 7, 10 *[1 mark]*

2. E.g. 288 ÷ –3 = –96
   –96 ÷ 12 = –8
   So the third number is –8
   [3 marks available — 1 mark for a correct method, 1 mark for at least one correct calculation, 1 mark for the correct answer]
   You could have worked out –3 × 12 (= –36) and divided by this instead.

3. –1.12, –0.61, –0.23, 0.35, 0.75, 1.06 *[1 mark]*

Answers

4   $(3 - -4) \times 5 = 7 \times 5 = 35$
    [3 marks available — 1 mark for (3 – –4), 1 mark for × 5,
    1 mark for correct answer]
    You might need to use trial and error for this one.

Page 9: Prime Numbers

1   29, 31, 37
    [2 marks available — 2 marks for all three prime numbers
    circled, 1 mark for two prime numbers circled. Lose 1 mark for
    each non-prime number circled (up to a maximum of 2 marks).]

2   a)   47 or 53 *[1 mark]*
    b)   67 or 71 *[1 mark]*

3   a)   7 or 11 *[1 mark]*
    b)   E.g. 7 and 12 *[1 mark]*
    There are a few correct answers here — 1 and 12 or 11 and 12
    are also correct.

4   Jack is incorrect as there are four prime numbers (101, 103, 107
    and 109) between 100 and 110.
    [2 marks available — 1 mark for stating that Jack is incorrect,
    1 mark for providing evidence]
    Writing any prime number between 100 and 110 is enough evidence.

5   E.g. 37 (3 + 7 = 10, which is 1 more than 9, a square number)
    [2 marks available — 2 marks for a correct answer, otherwise
    1 mark for a prime of two or more digits]

Page 10: Multiples, Factors and Prime Factors

1   a)   72 *[1 mark]*
    b)   64 *[1 mark]*
    c)   80 *[1 mark]*

2   a)   1, 2, 4, 7, 14, 28
        [2 marks available — 2 marks if all 6 factors are correct
        and no extra incorrect factors have been included,
        otherwise 1 mark if all 6 factors are correct but 1 extra
        incorrect factor has been included, or if at least 4 factors are
        correct and there are no more than 6 numbers listed in total]
    b)   56, 64 *[1 mark]*

3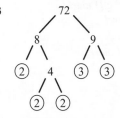

    $72 = 2 \times 2 \times 2 \times 3 \times 3$
    [2 marks available — 1 mark for a correct method,
    1 mark for all prime factors correct]

Page 11: LCM and HCF

1   a)   Multiples of 15 are: 15, 30, 45, <u>60</u>, 75, ...
        Multiples of 20 are: 20, 40, <u>60</u>, 80, ...
        So the lowest common multiple (LCM) is 60.
        [2 marks available — 1 mark for correct working,
        1 mark for the correct answer]
    b)   Factors of 42 are: 1, 2, 3, 6, 7, <u>14</u>, 21, 42
        Factors of 70 are: 1, 2, 5, 7, 10, <u>14</u>, 35, 70
        So the highest common factor (HCF) is 14.
        [2 marks available — 1 mark for correct working,
        1 mark for the correct answer]
        You <u>could</u> use the prime factors to go straight to finding
        the HCF, but there's a good chance of making a mistake.
        It's much safer to list <u>all</u> the factors and find the HCF
        that way, even if it takes a bit longer.

2   a)   LCM = $3^7 \times 7^3 \times 11^2$ *[1 mark]*
    b)   HCF = $3^4 \times 11$ *[1 mark]*

3   Multiples of 12 are:
    12, 24, 36, 48, 60, 72, 84, 96, 108, 120, 132, <u>144</u>, 156, ...
    Multiples of 16 are: 16, 32, 48, 64, 80, 96, 112, 128, <u>144</u>, 160, ...
    Multiples of 36 are: 36, 72, 108, <u>144</u>, 180, ...
    The LCM of 12, 16 and 36 is 144, which is the minimum number
    of each item he needs.
    The minimum number of packs of jars he needs is
    144 ÷ 12 = 12 packs
    The minimum number of packs of lids he needs is
    144 ÷ 16 = 9 packs
    The minimum number of packs of labels he needs is
    144 ÷ 36 = 4 packs
    [3 marks available — 1 mark for a correct method to find LCM,
    1 mark for LCM correct, 1 mark for all correct number of packs]

Pages 12-14: Fractions

1   a) $(60 \div 5) \times 3 = 12 \times 3 = 36$
        [2 marks available — 1 mark for dividing by 5 or
        multiplying by 3, 1 mark for the correct answer]
    b) $\frac{15}{40} = \frac{3}{8}$
        [2 marks available — 1 mark for putting the numbers into
        a fraction, 1 mark for the correct final answer]

2   a) $\frac{1}{2} \times \frac{1}{6} = \frac{1 \times 1}{2 \times 6} = \frac{1}{12}$ *[1 mark]*
    b) $\frac{2}{3} \div \frac{3}{5} = \frac{2}{3} \times \frac{5}{3} = \frac{2 \times 5}{3 \times 3} = \frac{10}{9}$ or $1\frac{1}{9}$
        [2 marks available — 1 mark for changing to the reciprocal
        fraction and multiplying, 1 mark for the correct answer]

3   $\frac{5}{6} = \frac{20}{24}$, $\frac{3}{4} = \frac{18}{24}$, $\frac{7}{8} = \frac{21}{24}$

    All these fractions are less than one, and the largest is $\frac{21}{24}$,
    so the fraction closest to 1 is $\frac{7}{8}$ *[1 mark]*

4   E.g. over half the shape would be shaded, but $\frac{3}{20}$
    is less than half *[1 mark]*.
    There are other reasons you could give here — for example,
    you could say that when adding the two fractions, Aito has added
    the denominators, which is incorrect.

5   $(12\,400 \div 8) \times 3 = 1550 \times 3 = 4650$
    [2 marks available — 1 mark for dividing by 8 or
    multiplying by 3, 1 mark for the correct answer]

6   $1 - \frac{2}{15} - \frac{5}{12} = \frac{60}{60} - \frac{8}{60} - \frac{25}{60} = \frac{27}{60} = \frac{9}{20}$
    [3 marks available — 1 mark for writing over a common
    denominator, 1 mark for 27/60, 1 mark for simplifying
    to find the correct answer]

7   Shaded regions are $\frac{1}{4}$, $\frac{1}{4} \times \frac{1}{4} = \frac{1}{16}$ and $\frac{1}{4} \times \frac{1}{4} \times \frac{1}{4} = \frac{1}{64}$
    So total area shaded = $\frac{1}{4} + \frac{1}{16} + \frac{1}{64} = \frac{16}{64} + \frac{4}{64} + \frac{1}{64} = \frac{21}{64}$
    [3 marks available — 1 mark for working out the fraction
    for each shaded region, 1 mark for writing over a common
    denominator, 1 mark for correct answer]

8   $17\frac{1}{2} \times \frac{1}{5} = \frac{35}{2} \times \frac{1}{5} = \frac{35}{10} = \frac{7}{2}$ *[1 mark]* tonnes of flour
    used to make cheese scones.
    Then $\frac{7}{2}$ out of 25 = $\frac{7}{2} \div 25 = \frac{7}{50}$ *[1 mark]*.
    [2 marks available in total — as above]

9   a) $1\frac{1}{8} \times 2\frac{2}{5} = \frac{9}{8} \times \frac{12}{5}$ *[1 mark]* = $\frac{108}{40}$ *[1 mark]*
        = $2\frac{7}{10}$ *[1 mark]*
        [3 marks available in total — as above]
    b) $1\frac{3}{4} \div \frac{7}{9} = \frac{7}{4} \times \frac{9}{7}$ *[1 mark]* = $\frac{63}{28}$ or $\frac{9}{4}$ *[1 mark]*
        = $2\frac{1}{4}$ *[1 mark]*
        [3 marks available in total — as above]

10 $5 \times \dfrac{5}{6} = \dfrac{25}{6} = 4\dfrac{1}{6}$ *[1 mark]*, so they will need 5 pizzas in total.
Cost = (2 × £7) + £4.50 *[1 mark]* = £14 + £4.50 = £18.50 *[1 mark]*
[3 marks available in total — as above]

11 Number of acres used for wheat = (36 ÷ 12) × 5 = 3 × 5 = 15
Number of acres used for cows = 36 ÷ 3 = 12
Number of acres used for pigs = 36 ÷ 6 = 6 *[1 mark for all 3 areas]*
Cost to run per year = (15 + 12 + 6) × £400 = 33 × £400
= £13 200 *[1 mark]*
Income from wheat = £1100 × 15 = £16 500
Income from cows = £1450 × 12 = £17 400
Income from pigs = £1250 × 6 = £7500 *[1 mark]*
Total profit = £16 500 + £17 400 + £7500 − £13 200 = £28 200
[1 mark]
[4 marks available in total — as above]
Alternatively you could work out the profit from each of wheat, cows and pigs, then add them up.

Page 15: Fractions, Decimals and Percentages

1 a) $\dfrac{3}{4} = 3 \div 4 = 0.75$ *[1 mark]*
   b) 0.06 × 100 = 6% *[1 mark]*

2 $\dfrac{6}{100}$ *[1 mark]*

3 65% = 0.65, $\dfrac{2}{3} = 0.666...$, $\dfrac{33}{50} = 0.66$
   So order is 0.065, 65%, $\dfrac{33}{50}$, $\dfrac{2}{3}$
   [2 marks available — 2 marks for all four numbers in the correct order, otherwise 1 mark for writing the numbers in the same form (either decimals, percentages or fractions)]

4 $\dfrac{1}{4} = 25\%$, so Jenny pays 1 − 25% − 20% − 20% = 1 − 65% = 35%
   [1 mark]
   £17.50 = 35% *[1 mark]*, so 1% = £17.50 ÷ 35 = £0.50.
   The total bill was £0.50 × 100 *[1 mark]* = £50 *[1 mark]*.
   [4 marks available in total — as above]

Page 16: Rounding

1 a) 120 *[1 mark]*
   b) 2600 *[1 mark]*
   c) 500 000 *[1 mark]*

2 a) 428.6 light years *[1 mark]*
   b) 430 light years *[1 mark]*

3 $\dfrac{4.32^2 - \sqrt{13.4}}{16.3 + 2.19} = 0.8113466...$ *[1 mark]*
   = 0.811 (3 s.f.) *[1 mark]*
   [2 marks available in total — as above]

4 Rounding unit = 1, so half of rounding unit = 1 ÷ 2 = 0.5
   Smallest possible value = 122 − 0.5 = 121.5 *[1 mark]*

Pages 17-18: Estimating and Error

1 E.g. Height of penguin ≈ 180 ÷ 3 *[1 mark]*
   = 60 cm (accept 50-67 cm) *[1 mark]*
   [2 marks available in total — as above]

2 a) E.g. (£4.95 × 28) + (£11 × 19) ≈ (£5 × 30) + (£10 × 20)
       = £150 + £200 = £350
       [2 marks available — 1 mark for rounding each value sensibly, 1 mark for a sensible estimate]
   b) E.g. This is a sensible estimate as it is very close to the actual value of £347.60 *[1 mark]*.

3 a) 1750 × 12 g ≈ 2000 × 10 = 20 000 g = 20 kg
       [2 marks available — 1 mark for rounding each value sensibly, 1 mark for a sensible estimate]
   b) E.g. Yes, an average person could lift approximately 20 kg of pens as this is a reasonable amount to lift *[1 mark]*.

4 E.g. $\dfrac{12.2 \times 1.86}{0.19} \approx \dfrac{10 \times 2}{0.2} = \dfrac{20}{0.2} = 100$
   [2 marks available — 1 mark for rounding to suitable values, 1 mark for the correct final answer using your values]

5 Minimum weight = 56.5 kg *[1 mark]*
   Maximum weight = 57.5 kg *[1 mark]*
   [2 marks available in total — as above]

6 a) Smallest possible value of a = 3.8 − 0.05 = 3.75
       Largest possible value of a = 3.8 + 0.05 = 3.85
       So error interval is 3.75 ≤ a < 3.85
       [2 marks available — 1 mark for 3.75 ≤ a, 1 mark for a < 3.85]
   b) Smallest possible value of b = 100.0 − 0.05 = 99.95
       Largest possible value of b = 100.0 + 0.05 = 100.05
       So error interval is 99.95 ≤ b < 100.05
       [2 marks available — 1 mark for 99.95 ≤ b, 1 mark for b < 100.05]

Page 19: Powers and Roots

1 a) $8.7^3 = 658.503$ *[1 mark]*
   b) $\sqrt[4]{1296} = 6$ *[1 mark]*
   c) $4^{-2} = 0.0625$ *[1 mark]*

2 $\sqrt{6.25} = 2.5$ cm *[1 mark]*

3 $\dfrac{3^4 \times 3^7}{3^6} = \dfrac{3^{(4+7)}}{3^6} = \dfrac{3^{11}}{3^6} = 3^{(11-6)} = 3^5$
   [2 marks available — 1 mark for a correct attempt at adding or subtracting powers, 1 mark for the correct final answer]

4 a) $6^{(5-3)} = 6^2 = 36$ *[1 mark]*
   b) $(2^4 \times 2^7) = 2^{(4+7)} = 2^{11}$
       $(2^3 \times 2^2) = 2^{(3+2)} = 2^5$, so $(2^3 \times 2^2)^2 = (2^5)^2 = 2^{10}$
       So $(2^4 \times 2^7) \div (2^3 \times 2^2)^2 = 2^{11} \div 2^{10} = 2^1 = 2$
       [2 marks available — 1 mark if each bracket has been correctly simplified, 1 mark for the correct answer]

Page 20: Standard Form

1 a) $A = 4.834 \times 10^9 = 4\ 834\ 000\ 000$ *[1 mark]*
   b) C, B, A (5.21 × 10^3, 2.4 × 10^5, 4.834 × 10^9) *[1 mark]*

2 a) Particle C *[1 mark]*
   b) $1.4 \times 10^{-6} = 0.0000014$ g *[1 mark]*
   c) $(3.2 \times 10^{-7}) - (2.1 \times 10^{-7}) = (3.2 - 2.1) \times 10^{-7}$ *[1 mark]*
       $= 1.1 \times 10^{-7}$ g *[1 mark]*
       [2 marks available in total — as above]

3 time (s) = distance (miles) ÷ speed (miles/s)
             = (9 × 10^7) ÷ (2 × 10^5) seconds *[1 mark]*
             = 450 seconds *[1 mark]*
   [2 marks available in total — as above]

Section Two — Algebra

Page 21: Algebra — Simplifying

1 10s *[1 mark]*

2 3x is a <u>term</u> in the <u>expression</u> $3x + 4y + 7$.
   [2 marks available — 1 mark for each correct word]

3 a) 4p *[1 mark]*
   b) 2m *[1 mark]*
   c) 4p + 3r
       [2 marks available — 1 mark for 4p and 1 mark for 3r]

4 a) 4pq *[1 mark]*
   b) $x^2 + 4x$
       [2 marks available — 1 mark for x^2 and 1 mark for 4x]

Pages 22-23: Algebra — Multiplying and Brackets

1 a) w^5 *[1 mark]*
   b) 10ab *[1 mark]*
   c) 2a *[1 mark]*

2 a) $3(x - 1) + 5(x + 2)$
$= 3x - 3 + 5x + 10$
$= 8x + 7$
[2 marks available — 1 mark for 8x, 1 mark for 7]

b) $4a(a + 2b)$
$= 4a^2 + 8ab$ *[1 mark]*

c) $9 - 3(x + 2)$
$= 9 - 3x - 6 = 3 - 3x$
[2 marks available — 1 mark for 3, 1 mark for –3x]

3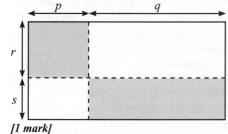

[1 mark]

4 $5(p + 6) - 2(p + 10) = 5p + 30 - 2p - 20 = 3p + 10$
[2 marks available — 1 mark for expanding both brackets correctly, 1 mark for simplifying to 3p + 10]

5 a) $(x + 2)(x + 4) = x^2 + 4x + 2x + 8 = x^2 + 6x + 8$
[2 marks available — 1 mark for expanding the brackets correctly, 1 mark for simplifying]

b) $(y + 3)(y - 3) = y^2 - 3y + 3y - 9 = y^2 - 9$
[2 marks available — 1 mark for expanding the brackets correctly, 1 mark for simplifying]

c) $(2z - 1)(z - 5) = 2z^2 - 10z - z + 5 = 2z^2 - 11z + 5$
[2 marks available — 1 mark for expanding the brackets correctly, 1 mark for simplifying]

6 a) $(a - 7)^2 = (a - 7)(a - 7) = a^2 - 7a - 7a + 49 = a^2 - 14a + 49$
[2 marks available — 1 mark for expanding the brackets correctly, 1 mark for simplifying]

b) $(3b + 2)^2 = (3b + 2)(3b + 2) = 9b^2 + 6b + 6b + 4 = 9b^2 + 12b + 4$
[2 marks available — 1 mark for expanding the brackets correctly, 1 mark for simplifying]

Page 24: Factorising

1 $6x + 3 = (3 \times 2x) + (3 \times 1) = 3(2x + 1)$ *[1 mark]*

2 a) $7y - 21y^2 = 7(y - 3y^2)$
$= 7y(1 - 3y)$
[2 marks available — 2 marks for the correct final answer, otherwise 1 mark if the expression is only partly factorised]

b) $4x^2 + 6xy = 2(2x^2 + 3xy)$
$= 2x(2x + 3y)$
[2 marks available — 2 marks for the correct final answer, otherwise 1 mark if the expression is only partly factorised]

3 a) $x^2 - 49 = x^2 - 7^2 = (x + 7)(x - 7)$
[2 marks available — 2 marks for the correct final answer, otherwise 1 mark for attempting to use the difference of two squares]

b) $9x^2 - 100 = (3x)^2 - 10^2 = (3x + 10)(3x - 10)$
[2 marks available — 2 marks for the correct final answer, otherwise 1 mark for attempting to use the difference of two squares]

c) $y^2 - m^2 = (y + m)(y - m)$
[2 marks available — 2 marks for the correct final answer, otherwise 1 mark for attempting to use the difference of two squares]

Pages 25-26: Solving Equations

1 a) $x + 3 = 12$
$x = 9$ *[1 mark]*

b) $6x = 24$
$x = 4$ *[1 mark]*

c) $\frac{x}{5} = 4$
$x = 20$ *[1 mark]*

2 a) $p - 11 = -7$
$p = 4$ *[1 mark]*

b) $2y - 5 = 9$
$2y = 14$ *[1 mark]*
$y = 7$ *[1 mark]*
[2 marks available in total — as above]

c) $3z + 2 = z + 15$
$2z = 13$ *[1 mark]*
$z = 13 \div 2 = 6.5$ *[1 mark]*
[2 marks available in total — as above]

3 a) $40 - 3x = 17x$
$40 = 20x$ *[1 mark]*
$x = 40 \div 20 = 2$ *[1 mark]*
[2 marks available in total — as above]

b) $2y - 5 = 3y - 12$
$-5 + 12 = 3y - 2y$ *[1 mark]*
$y = 7$ *[1 mark]*
[2 marks available in total — as above]

4 a) $3(a + 2) = 15$
$3a + 6 = 15$ *[1 mark]*
$3a = 9$ *[1 mark]*
$a = 3$ *[1 mark]*
[3 marks available in total — as above]

b) $5(2b - 1) = 4(3b - 2)$
$10b - 5 = 12b - 8$ *[1 mark]*
$3 = 2b$ *[1 mark]*
$b = 1.5$ *[1 mark]*
[3 marks available in total — as above]

5 $(x + 2)(x - 4) = (x - 2)(x + 1)$
$x^2 - 2x - 8 = x^2 - x - 2$
$-8 + 2 = -x + 2x$
$-6 = x$ so $x = -6$
[4 marks available — 1 mark for expanding the brackets on the RHS, 1 mark for expanding the brackets on the LHS, 1 mark for collecting like terms on each side, 1 mark for the correct solution]

6 $6w^2 = 600$
$w^2 = 100$ *[1 mark]*
$w = \pm\sqrt{100}$ *[1 mark]*
$w = \pm 10$ *[1 mark]*
[3 marks available in total — as above]

Page 27: Expressions, Formulas and Functions

1 a) $S = 4m^2 + 2.5n$
$S = (4 \times 2 \times 2) + (2.5 \times 10)$
$S = 16 + 25 = 41$
[2 marks available — 1 mark for correct substitution of m and n, 1 mark for correct final answer]

b) $S = 4m^2 + 2.5n$
$S = (4 \times 6.5 \times 6.5) + (2.5 \times 4)$
$S = 169 + 10 = 179$
[2 marks available — 1 mark for correct substitution of m and n, 1 mark for correct final answer]

2 a) an expression *[1 mark]*

b) an equation *[1 mark]*

3 a) $23 + 7 = 30$
$30 \div 5 = 6$, so when $x = 23$, $y = 6$ *[1 mark]*

b) $3 \times 5 = 15$
$15 - 7 = 8$, so when $y = 3$, $x = 8$
[2 marks available — 1 mark for reversing the function machine, 1 mark for the correct value of x]

Pages 28-30: Equations from Words and Diagrams

1 a) Number of miles = (number of kilometres ÷ 8) × 5
   $m = (k \div 8) \times 5$
   $m = \dfrac{5k}{8}$
   [2 marks available — 2 marks for correct formula,
   otherwise 1 mark for just $\dfrac{5k}{8}$]

   b) Substitute $k = 110$ into formula:
   $m = \dfrac{5 \times 110}{8}$
   $m = 550 \div 8 = 68.75$
   Therefore 110 km = 68.75 miles.
   [2 marks available — 1 mark for substitution of k = 110 into
   formula, 1 mark for correct final answer]

2 Call the number of cakes Nancy bakes n. Then Chetna bakes $2n$
   cakes and Norman bakes $2n + 12$ cakes. They bake 72 cakes, so
   $n + 2n + 2n + 12 = 72$
   $5n + 12 = 72$
   $5n = 60$
   $n = 12$
   So Nancy bakes 12 cakes, Chetna bakes $12 \times 2 = 24$ cakes
   and Norman bakes $24 + 12 = 36$ cakes.
   [4 marks available — 1 mark for forming expressions for the
   number of cakes each person bakes, 1 mark for forming an
   equation for the total number of cakes baked, 1 mark for solving
   the equation, 1 mark for the correct numbers of cakes each
   person bakes]

3 a) $P = (4x - 3) + (x - 4) + (4x - 3) + (x - 4)$
   $= 10x - 14$
   [2 marks available — 1 mark for adding up the side lengths,
   1 mark for simplifying the expression for P]

   b) $10x - 14 = 36$ *[1 mark]*
   $10x = 50$
   $x = 5$ *[1 mark]*
   [2 marks available in total — as above]

4 a) Call Jessica's number j. Then $j^2 - 7 = 57$ *[1 mark]*,
   so $j^2 = 64$, which means $j = 8$ *[1 mark]*.
   [2 marks available in total — as above]
   You can ignore the negative square root as you're told
   that her number is positive.

   b) Call Ricardo's number r. Then $\sqrt{r} + 13 = 18$ *[1 mark]*,
   so $\sqrt{r} = 5$, which means $r = 5^2 = 25$ *[1 mark]*.
   [2 marks available in total — as above]

5 The sides of an equilateral triangle are all the same length, so
   $4(x - 1) = 3x + 5$ *[1 mark]*
   $4x - 4 = 3x + 5$
   $x = 9$ *[1 mark]*
   So each side is $(3 \times 9) + 5 = 32$ cm long *[1 mark]*.
   [3 marks available in total — as above]
   To check your answer, put your value of x into the expression for the
   other side of the triangle — you should get the same answer.

6 Call Peter's time t minutes. Cassie's time = $(t + 2)$ minutes and
   Lisa's time = $(t - 4)$ minutes.
   Total time = $t + (t + 2) + (t - 4) = (3t - 2)$ minutes.
   So $3t - 2 = 43$, then $3t = 45$, so $t = 15$. So Peter takes 15 minutes,
   Cassie takes $15 + 2 = 17$ minutes and Lisa takes $15 - 4 = 11$ minutes.
   [3 marks available — 1 mark for forming the equation for the
   total time, 1 mark for solving the equation to find t, 1 mark for
   the correct answer]

7 Call the number of Whitewater fans f. Redwood fans = $3 \times f = 3f$.
   Difference = $3f - f = 2f$, so $2f = 7000$, so $f = 3500$.
   Total fans = $3f + f = 4f = 4 \times 3500 = 14\,000$.
   [3 marks available — 1 mark for the expressions for the number
   of fans for each team, 1 mark for forming and solving the
   equation to find f, 1 mark for the correct answer]

8 Perimeter of triangle = $(4x - 7) + (4x - 7) + (2x - 3) = 10x - 17$ cm
   Perimeter of square = $4 \times (x + 2) = 4x + 8$ cm
   So $10x - 17 = 2(4x + 8)$
   $10x - 17 = 8x + 16$
   $2x = 33$
   $x = 16.5$ cm
   So the base of the isosceles triangle is $(2 \times 16.5) - 3 = 30$ cm.
   [4 marks available — 1 mark for the expressions for the
   perimeters of the triangle and square, 1 mark for setting the
   triangle's perimeter equal to double the square's, 1 mark for
   solving the equation to find x, 1 mark for the correct answer]

9 Call the largest number x and the smallest number y.
   Then the middle number = $4 \times y = 4y$.
   The sum of the three numbers is $x + y + 4y = x + 5y$,
   so $x + 5y = 25$.
   Now find x: multiples of 4 = 4, 8, 12, 16, 20, 24
   $25 - x$ needs to give a multiple of 5, so $x = 20$.
   Then $20 + 5y = 25$, so $5y = 5$ which means $y = 1$
   So the three numbers are 1, 4 and 20.
   [3 marks available — 1 mark for each of the following criteria:
   one number that is a multiple of four, a pair of numbers where
   one number is four times the other, three numbers that add up to
   twenty-five]
   You could have solved this one using trial and error.

Page 31: Rearranging Formulas

1 a) $u = v - at$ *[1 mark]*
   b) $v - u = at$ *[1 mark]*
   $t = \dfrac{v - u}{a}$ *[1 mark]*
   [2 marks available in total — as above]

2 $\dfrac{a + 2}{3} = b - 1$
   $a + 2 = 3b - 3$ *[1 mark]*
   $a = 3b - 5$ *[1 mark]*
   [2 marks available in total — as above]

3 $x = y^2 - 7$
   $x + 7 = y^2$ *[1 mark]*
   $y = \pm\sqrt{x + 7}$ *[1 mark]*
   [2 marks available in total — as above]

4 $u = 2 + \dfrac{1}{w}$
   $u - 2 = \dfrac{1}{w}$ *[1 mark]*
   $w(u - 2) = 1$ *[1 mark]*
   $w = \dfrac{1}{u - 2}$ *[1 mark]*
   [3 marks available in total — as above]

Pages 32-33: Sequences

1 a) 36, 44 *[1 mark]*
   b) 23rd term = 25th term − (2 × difference between terms)
   $= 196 - (2 \times 8) = 196 - 16 = 180$ *[1 mark]*
   c) All terms in the sequence must be a multiple of 4 *[1 mark]*
   (the first term is 4, and the difference between the terms is 8).
   90 isn't a multiple of 4, so it can't be the 12th term. *[1 mark]*
   [2 marks available in total — as above]
   You could also work out the 12th term and show that it's not 90.

2 a)
   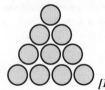 *[1 mark]*
   b) The number of circles added increases by one each time,
   so the tenth triangle number is:
   $1 + 2 + 3 + 4 + 5 + 6 + 7 + 8 + 9 + 10 = 55$.
   [2 marks available — 1 mark for 55 and 1 mark for correct
   reasoning]

3 Second term = 7 – 3 = 4
   Fourth term = 4 + 7 = 11
   Fifth term = 7 + 11 = 18
   [2 marks available — 2 marks for all three terms correct,
   otherwise 1 mark for at least one term correct]

4 a) 2   9   16   23
         +7 +7   +7

      The common difference is 7, so 7n is in the formula.
          n = 1   2   3   4
         7n = 7   14   21   28
              |–5 |–5 |–5 |–5
      nth term = 2   9   16   23
      You have to subtract 5 to get to the term.
      So the expression for the nth term is 7n – 5.
      [2 marks available — 2 marks for the correct expression,
      otherwise 1 mark for 7n]

   b) 30th term = (7 × 30) – 5 = 205 *[1 mark]*

   c) If 55 is a term in the sequence, then 7n – 5 = 55
      7n = 60 so n = 8.571...
      n is not a whole number, so 55 is not a term in the sequence.
      [2 marks available — 1 mark for the correct answer,
      1 mark for a suitable explanation]

5 2   6   12   20
     +4 +6   +8

   The difference is increasing by 2, so the next term is: 20 + 10 = 30
   [2 marks available — 1 mark for spotting the pattern, 1 mark for
   the correct answer]

6 Two consecutive terms are nth and (n + 1)th, which have values:
   3n – 10 and 3(n + 1) – 10 = 3n – 7.
   Their sum is 3n – 10 + 3n – 7 = 6n – 17.
   So 6n – 17 = 1
   6n = 18 and n = 3
   So the two terms are (3 × 3) – 10 = –1 and (3 × 3) – 7 = 2.
   [4 marks available — 1 mark for finding expressions for both
   terms, 1 mark for setting their sum equal to 1, 1 mark for solving
   the equation, 1 mark for both correct terms]

Page 34: Inequalities

1 $x \geq -2$ *[1 mark]*
   It's ≥ because the circle above the number line is coloured in,
   so –2 is included.

2 –3, –2, –1, 0, 1
   [2 marks available — 2 marks for all 5 numbers correct,
   otherwise 1 mark for the correct answer with one number
   missing or one number incorrect]

3 Largest possible value of p = 45
   Smallest possible value of q = 26 *[1 mark for both]*
   Largest possible value of p – q = 45 – 26 = 19 *[1 mark]*.
   [2 marks available in total — as above]

4 a) 2p > 4
      p > 4 ÷ 2
      p > 2 *[1 mark]*
   b) 4q – 5 < 23
      4q < 23 + 5
      4q < 28 *[1 mark]*
      q < 28 ÷ 4
      q < 7 *[1 mark]*
      [2 marks available in total — as above]
   c) 4r – 2 ≥ 6r + 5
      4r – 6r ≥ 5 + 2
      –2r ≥ 7 *[1 mark]*
      r ≤ 7 ÷ –2
      r ≤ –3.5 *[1 mark]*
      [2 marks available in total — as above]

Page 35: Quadratic Equations

1 3 and 6 multiply to give 18 and add to give 9,
   so $x^2 + 9x + 18 = (x + 3)(x + 6)$
   [2 marks available — 1 mark for correct numbers in brackets,
   1 mark for correct signs]
   The brackets can be either way around —
   (x + 6)(x + 3) is also correct.

2 1 and 5 multiply to give 5 and subtract to give –4,
   so $y^2 – 4y – 5 = (y + 1)(y – 5)$
   [2 marks available — 1 mark for correct numbers in brackets,
   1 mark for correct signs]

3 4 and 8 multiply to give 32 and subtract to give 4,
   so $x^2 + 4x – 32 = (x – 4)(x + 8)$
   [2 marks available — 1 mark for correct numbers in brackets,
   1 mark for correct signs]

4 a) 4 and 5 multiply to give 20 and add to give 9,
      so $x^2 – 9x + 20 = (x – 4)(x – 5)$
      [2 marks available — 1 mark for correct numbers in brackets,
      1 mark for correct signs]
   b) x – 4 = 0   or   x – 5 = 0
      x = 4   or   x = 5
      [1 mark for both solutions correct]

5 6 and 2 multiply to give 12 and subtract to give 4,
   so if $x^2 + 4x – 12 = 0$,
   $(x + 6)(x – 2) = 0$
   [1 mark for correct numbers in brackets,
   1 mark for correct signs]
   x + 6 = 0   or   x – 2 = 0
   x = –6   or   x = 2
   [1 mark for both solutions]
   [3 marks available in total — as above]

Page 36: Simultaneous Equations

1 4x + 3y = 16 (1)
   4x + 2y = 12 (2)
   (1) – (2):
   4x + 3y = 16   4x + 3y = 16
   – 4x + 2y = 12   4x + 12 = 16
      y = 4 *[1 mark]*   x = 1 *[1 mark]*
   [2 marks available in total — as above]

2 3x + 4y = 26 (1)
   2x + 2y = 14 (2) —×2→ 4x + 4y = 28 (3) *[1 mark]*
   (3) – (1):
   4x + 4y = 28   3x + 4y = 26
   – 3x + 4y = 26   (3 × 2) + 4y = 26
      x = 2 *[1 mark]*   4y = 26 – 6 = 20
                             y = 5 *[1 mark]*
   [3 marks available in total — as above]

3 x + 3y = 11 (1) —×3→ 3x + 9y = 33 (3) *[1 mark]*
   3x + y = 9 (2)
   (3) – (2):
   3x + 9y = 33   x + 3y = 11
   – 3x + y = 9   x + (3 × 3) = 11
      8y = 24   x = 11 – 9
      y = 3 *[1 mark]*   x = 2 *[1 mark]*
   [3 marks available in total — as above]

4 2x + 3y = 12 (1) —×5→ 10x + 15y = 60 (3) *[1 mark]*
   5x + 4y = 9 (2) —×2→ 10x + 8y = 18 (4) *[1 mark]*
   (3) – (4):
   10x + 15y = 60   2x + 3y = 12
   – 10x + 8y = 18   2x = 12 – (3 × 6)
      7y = 42   2x = –6
      y = 6 *[1 mark]*   x = –3 *[1 mark]*
   [4 marks available in total — as above]

Page 37: Proof

1 a) 16 is a factor of 48 *[1 mark]*

b) E.g. 4 + 16 = 20, which is even *[1 mark]*

c) E.g. 38 is not a multiple of 4, 6 or 8 *[1 mark]*

2 E.g. 5 × 3 = 15, which is not a multiple of 9 *[1 mark]*

3 LHS: $(x + 2)^2 + (x - 2)^2 = x^2 + 4x + 4 + x^2 - 4x + 4$ *[1 mark]*
$= 2x^2 + 8$ *[1 mark]*
$= 2(x^2 + 4)$ = RHS *[1 mark]*

[3 marks available in total — as above]

4 $2(18 + 3q) + 3(3 + q) = 36 + 6q + 9 + 3q$
$= 9q + 45$
$= 9(q + 45)$

$2(18 + 3q) + 3(3 + q)$ can be written as 9 × a whole number (where the whole number is $(q + 45)$), so it is a multiple of 9.

[3 marks available — 1 mark for expanding brackets and simplifying, 1 mark for writing the expression as 9(q + 45), 1 mark for explaining why this is a multiple of 9]

Section Three — Graphs

Page 38: Coordinates and Midpoints

1 a) (2, 1) *[1 mark]*

b) (3, –2) *[1 mark]*

c)

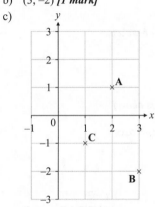

[1 mark]

2 a) $\left(\dfrac{1 + 3}{2}, \dfrac{3 + (-1)}{2}\right) = (2, 1)$

[2 marks available — 1 mark for correct method and 1 mark for correct final answer]

A correct method here is to find the averages of the x- and y-coordinates. Or, you could identify the midpoint of AB on the graph to get your answer — but the first way is much safer.

b) Comparing coordinates of point **C** and midpoint of **CD**:
x-distance = 2 – 0 = 2
y-distance = 1 – –1 = 2
So to get from the midpoint to point **D**, move up 2 and right 2.
So point **D** is (2 + 2, 1 + 2) = (4, 3)

[2 marks available — 1 mark for each correct coordinate]

Pages 39-41: Straight-Line Graphs

1 a)

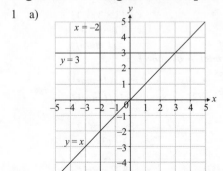

[3 marks available — 1 mark for each correct line]

b) (3, 3) *[1 mark]*

2 a)

x	–2	–1	0	1	2
y	–8	–5	–2	1	**4**

[2 marks available — 2 marks for all values correct, otherwise 1 mark for 2 correct values]

b)

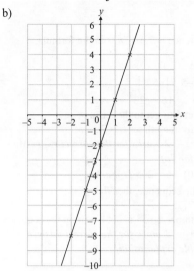

[2 marks available — 2 marks for all points plotted correctly and a straight line drawn from (–2, –8) to (2, 4), otherwise 1 mark for a correct straight line that passes through at least 3 correct points, or a straight line with the correct gradient, or a straight line with a positive gradient passing through (0, –2)]

c)

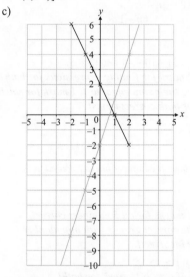

[3 marks available — 3 marks for a correct line drawn from (–2, 6) to (2, –2), otherwise 2 marks for a line that passes through (0, 2) and has a gradient of –2, or 1 mark for a line passing through (0, 2), or a line with a gradient of –2]

3 a) –2 *[1 mark]*

b) 1 *[1 mark]*

c) $\dfrac{1}{2}$ *[1 mark]*

4 Find the gradient: $\dfrac{\text{change in } y}{\text{change in } x} = \dfrac{4 - 1}{1 - 0} = 3$

Line crosses y-axis at 1, so equation of line is $y = 3x + 1$.

[3 marks available — 3 marks for a fully correct answer, otherwise 2 marks for a correct gradient, or 1 mark for a correct method to find the gradient.]

5 a)

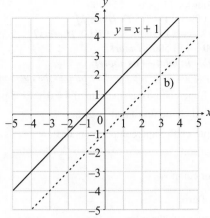

[2 marks available — 1 mark for correct gradient, 1 mark for correct intersection with y-axis]

b) Draw line parallel to $y = x + 1$ that passes through $(2, 1)$, — see dashed line on grid above.

m = 1 and c = –1, so $y = x – 1$

[2 marks available — 1 mark for correct line on graph, 1 mark for correct equation]

6 The lines are parallel, so their gradients are equal: m = 4 *[1 mark]*

When $x = -1$, $y = 0$, so put this into $y = 4x + c$

$0 = (4 \times -1) + c$, so c = 4 *[1 mark]*

So equation of line is $y = 4x + 4$ *[1 mark]*

[3 marks available in total — as above]

7 a) Gradient = $\dfrac{\text{Change in } y}{\text{Change in } x} = \dfrac{7 - (-1)}{5 - 1} = 2$

Use point A to find c:

So $y = 2x + c$

$-1 = (2 \times 1) + c$

$c = -3$

So $y = 2x - 3$

[4 marks available — 1 mark for a correct method for finding the gradient, 1 mark for correct gradient, 1 mark for putting one point into the equation, 1 mark for correct answer]

b) m = $\dfrac{6 - 0}{4 - 0}$

m = 1.5 *[1 mark]*

This gradient is different to the gradient of line **L**, so the two lines can't be parallel. *[1 mark]*

[2 marks available in total — as above]

Pages 42-43: Quadratic and Harder Graphs

1 a) 0 *[1 mark]*

b) –2, 0 *[1 mark]*

c) (–1, –1) *[1 mark]*

2 a)

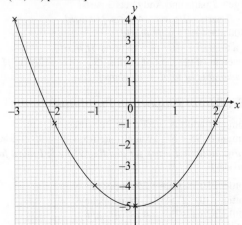

[2 marks available — 1 mark if all points are plotted correctly, 1 mark for a smooth curve joining the correctly plotted points]

b) $x = -2.2$ (allow –2.3 to –2.1) *[1 mark]*

3 a)

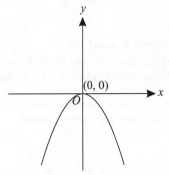

[2 marks available — 1 mark for correct shape, 1 mark for labelling (0, 0).]

b)

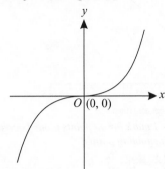

[2 marks available — 1 mark for correct shape, 1 mark for labelling (0, 0)]

4 A *[1 mark]*

Page 44: Solving Equations Using Graphs

1 $x = 3$ and $y = 4$ *[1 mark]*

These are the x and y coordinates of the point where the two lines cross.

2 a) $x = 1$ *[1 mark]*

b)

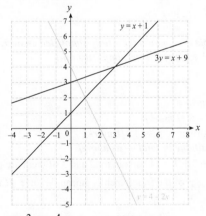

$x = 3$, $y = 4$

[3 marks available — 2 marks for correctly drawing the line 3y = x + 9, 1 mark for the correct answer]

Page 45: Distance-Time Graphs

1 a) 1 hour *[1 mark]*

b) Tyrone. He reaches 30 km after 5 hours whereas Selby reaches 30 km after 6 hours. *[1 mark]*

c) Gradient = $\dfrac{\text{change in } y}{\text{change in } x} = \dfrac{25 - 15}{3 - 1.5} = \dfrac{10}{1.5} = 6.67$ km/h (2 d.p)

[2 marks available — 2 marks for correct answer, otherwise 1 mark for choosing correct x and y values]

d) E.g. Selby is the most likely to have been injured. The gradient of Selby's line decreases towards the end of the race, whereas Tyrone's gets much steeper. This means Selby was moving much more slowly than Tyrone towards the end of the race.

[2 marks available — 1 mark for stating Selby is the injured runner, 1 mark for a correct explanation referring to gradients or steepness of lines]

Page 46: Real-Life Graphs

1 a) 5.5 gallons *[1 mark]*

b) 47 litres (allow 46-48 litres) *[1 mark]*

c) (i) E.g. Find 40 litres in gallons, and then double the answer.
(ii) 40 litres ≈ 8.8 gallons, so 80 litres ≈ 8.8 × 2 = 17.6 gallons
(allow 17.2-18.0 gallons)
*[2 marks available — 1 mark for a correct method to find
80 litres in gallons, 1 mark for a correct value]*

2 a)

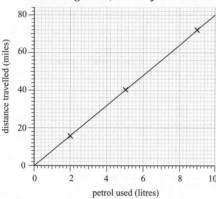

*[2 marks available — 1 mark for all points plotted correctly,
1 mark for straight line joining points]*

b) Gradient = $\dfrac{\text{change in } y}{\text{change in } x} = \dfrac{80 - 0}{10 - 0} = 8$

*[2 marks available — 1 mark for correct method to find
gradient, 1 mark for correct answer]*

c) Distance travelled in miles per litre of petrol used *[1 mark]*

Section Four —
Ratio, Proportion and Rates of Change

Pages 47-49: Ratios

1   $16:240 = (16 \div 8):(240 \div 8)$
$= 2:30 = 1:15$
*[2 marks available — 2 marks for a fully simplified answer,
otherwise 1 mark for any correct simplification]*

2 a) boys:girls
$= 12:14$
$=   6:7$ *[1 mark]*

b) $25 \div (2 + 3) = 5$ *[1 mark]*
Number of girls is $5 \times 3 = 15$ *[1 mark]*
[2 marks available in total — as above]

3 a) There are $4 + 3 + 7 = 14$ parts in total and 3 of them are pineapple juice. $\frac{3}{14}$ of the fruit punch is pineapple juice. *[1 mark]*

b) $700 \div (4 + 3 + 7) = 700 \div 14$
$= 50$ ml per part
Apple juice: $50 \times 4 = 200$ ml
Pineapple juice : $50 \times 3 = 150$ ml
Cherryade: $50 \times 7 = 350$ ml
*[3 marks available — 1 mark for dividing 700 by the sum of
the numbers in the ratio, 1 mark for multiplying this value by
each number in the ratio, 1 mark if all three quantities are
correct]*
*You might have worked this out using fractions — this method is
fine, but check that your final answers match those in this solution.*

4 a) Shortest side of shape A = 3 units
Shortest side of shape B = 6 units
Ratio of shortest sides = $3:6 = 1:2$
*[2 marks available — 1 mark for finding the shortest sides of
the triangles, 1 mark for the correct answer]*

b) Area of shape A = $\frac{1}{2} \times 3 \times 4 = 6$ square units *[1 mark]*
Area of shape B = $\frac{1}{2} \times 6 \times 8 = 24$ square units *[1 mark]*
Ratio of areas = $6:24 = 1:4$ *[1 mark]*
[3 marks available in total — as above]

5   Donations account for 14 parts = £21 000
So 1 part = £21 000 ÷ 14 = £1500 *[1 mark]*
Bills are 5 parts so cost £1500 × 5 = £7500 *[1 mark]*
£21 000 – £7500 = £13 500 *[1 mark]*
[3 marks available in total — as above]
Careful here — you are given a part:whole ratio in the question.

6   Mr Appleseed's Supercompost is made up of $4 + 3 + 1 = 8$ parts,
so contains: $\frac{4}{8}$ soil, $\frac{3}{8}$ compost and $\frac{1}{8}$ grit.
16 kg of Mr Appleseed's Supercompost contains:
$\frac{4}{8} \times 16 = 8$ kg of soil
$\frac{3}{8} \times 16 = 6$ kg of compost
$\frac{1}{8} \times 16 = 2$ kg of grit
Soil costs £8 ÷ 40 = £0.20 per kg.
Compost costs £15 ÷ 25 = £0.60 per kg.
Grit costs £12 ÷ 15 = £0.80 per kg.
16 kg of Mr Appleseed's Supercompost costs:
$(8 \times 0.2) + (6 \times 0.6) + (2 \times 0.8) = £6.80$
*[5 marks available — 1 mark for finding the fractions of each
material in the mix, 1 mark for the correct mass of one material,
1 mark for the correct masses for the other two materials,
1 mark for working out the price per kg for each material,
1 mark for the correct answer]*

7 a)

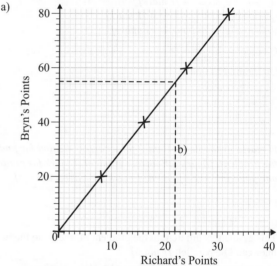

*[2 marks available — 1 mark for two points marked correctly,
1 mark for the correct straight line]*
Use the ratio to work out the coordinates of a few points to plot.
E.g. If Richard scored 8 points, Bryn scored $8 \times \frac{5}{2} = 20$ points.

b) 55 points (see graph) *[1 mark]*

8   Christine gets 7 parts and Andy gets 3 parts.
So 300 g = 7 parts – 3 parts = 4 parts *[1 mark]*
1 part = 300 g ÷ 4 = 75 g *[1 mark]*
There are a total of $3 + 6 + 7 = 16$ parts *[1 mark]*
The total weight of the joint of beef is 75 g × 16 = 1200 g *[1 mark]*
[4 marks available in total — as above]

Pages 50-51: Direct Proportion Problems

1   A minimum of 1 adult is needed per 5 children.
So a minimum of 95 ÷ 5 = 19 adults are needed for 95 children.
*[2 marks available — 1 mark for dividing 95 by 5, 1 mark for the
correct answer]*

2   250 ml bottle: 250 ÷ 200 = 1.25 ml per penny
330 ml bottle: 330 ÷ 275 = 1.2 ml per penny
525 ml bottle: 525 ÷ 375 = 1.4 ml per penny
So the 525 ml bottle is the best value for money.
*[3 marks available — 3 marks for finding the correct amounts
per penny for all three bottles and the correct answer, otherwise
2 marks for two correct amounts per penny or 1 mark for one
correct amount per penny]*
You could also compare the cost per ml of each bottle.

3   She worked 28 hours and got £231 so she gets paid
£231 ÷ 28 = £8.25 per hour *[1 mark]*.
So in total, she'll get paid
£231 + (3 × 25 × £8.25) = £849.75 *[1 mark]*.
[2 marks available in total — as above]

4   1 bottle of water costs £52.50 ÷ 42 = £1.25 *[1 mark]*.
There are £35 ÷ £1.25 = 28 girls in the club *[1 mark]*
[2 marks available in total — as above]

5   a)   For 1 sponge cake she'd need:
      Flour: 275 g ÷ 5 = 55 g   Butter: 275 g ÷ 5 = 55 g
      Sugar: 220 g ÷ 5 = 44 g   Eggs: 5 ÷ 5 = 1 egg
      So for 18 sponge cakes she'll use:
      Flour: 55 g × 18 = 990 g   Butter: 55 g × 18 = 990 g
      Sugar: 44 g × 18 = 792 g   Eggs: 1 × 18 = 18 eggs
      [3 marks available — 1 mark for dividing the quantities by 5,
      1 mark for multiplying the quantities by 18, 1 mark for all
      four correct answers]
      An alternative method is to find the multiplier from 5 to 18
      (it's 18 ÷ 5 = 3.6) and multiply each quantity in the original
      recipe by this number.

   b)   There will be a total of 18 × 10 = 180 slices *[1 mark]*
      At 50p each this will make:
      180 × 50p = 9000p = £90 *[1 mark]*
      Profit = £90 − £25.30 = £64.70 *[1 mark]*
      [3 marks available in total — as above]
      Any correct method showing full working and leading to the
      correct answer will get full marks.

6   a)   42.5 m² is enough for a room with a perimeter of 17 m.
      1 m² of wallpaper is enough for a room with a perimeter of
      17 ÷ 42.5 = 0.4 m
      55 m² of wallpaper is enough for a room with a perimeter of
      0.4 × 55 = 22 m
      [2 marks available — 1 mark for a correct method, 1 mark
      for the correct answer]

   b)

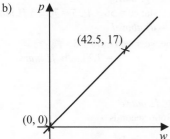

      [3 marks available — 1 mark for a straight line, 1 mark for a
      line going through the origin, 1 mark for marking two correct
      points (one can be the origin)]

Page 52: Inverse Proportion Problems

1   12 people take 3 hours.
1 person will take 3 × 12 = 36 hours.
4 people will take 36 ÷ 4 = 9 hours.
[2 marks available — 1 mark for a correct method, 1 mark for the
correct answer]
Alternatively, there are a third of the people (12 ÷ 4 = 3) so it will
take three times as long — 3 × 3 = 9 hours.

2   a)   250 people can be catered for 6 days
      1 person can be catered for 6 × 250 = 1500 days
      300 people can be catered for 1500 ÷ 300 = 5 days
      [2 marks available — 1 mark for a correct method,
      1 mark for the correct answer]

   b)   For a 1-day cruise it could cater for 6 × 250 = 1500 people
      For a 2-day cruise it could cater for 1500 ÷ 2 = 750 people
      So it can cater for 750 − 250 = 500 more people
      [3 marks available — 1 mark for a correct method to find the
      number of people catered for on a 2-day cruise, 1 mark for
      the correct number of people catered for on a 2-day cruise,
      1 mark for the correct final answer]

3   You need to be able to write the equations in the form $f = \dfrac{A}{g}$
where A is a number.
$fg = 7$ can be rearranged to $f = \dfrac{7}{g}$
$g = \dfrac{3}{f}$ can be rearranged to $f = \dfrac{3}{g}$
[2 marks available — 1 mark for circling each of these equations]

4   1 litre of petrol will keep 8 go-karts going for
24 ÷ 12 = 2 minutes *[1 mark]*
18 litres of petrol will keep 8 go-karts going for
2 × 18 = 36 minutes *[1 mark]*
18 litres of petrol will keep 1 go-kart going for
36 × 8 = 288 minutes *[1 mark]*
18 litres of petrol will keep 6 go-karts going for
288 ÷ 6 = 48 minutes *[1 mark]*
[4 marks available in total — as above]

Pages 53-54: Percentages

1   a)   10% of £18 = £18 ÷ 10 = £1.80
      [2 marks available — 1 mark for a correct method,
      1 mark for the correct answer]

   b) $\dfrac{6}{24} \times 100 = \dfrac{1}{4} \times 100 = 25\%$
      [2 marks available — 1 mark for a correct method,
      1 mark for the correct answer]

2   100% of 5200 = 5200
10% of 5200 = 5200 ÷ 10 = 520
5% of 5200 = 520 ÷ 2 = 260
115% = 5200 + 520 + 260 = 5980
[2 marks available — 1 mark for a correct method, 1 mark for the
correct answer]

3   20% increase = 1 + 0.2 = 1.2
20% increase of £33.25 = 1.2 × £33.25 = £39.90
[2 marks available — 1 mark for a correct method,
1 mark for the correct answer]

4   He normally gets 240 ÷ 40 = 6 packs *[1 mark]*
40% cheaper = 1 − 0.4 = 0.6
So the stickers are 40p × 0.6 = 24p per pack this week *[1 mark]*
He can buy 240 ÷ 24 = 10 packs this week *[1 mark]*
So he can get 10 − 6 = 4 more packs *[1 mark]*
[4 marks available in total — as above]

5   261 + 185 + 154 − 100 = 500 items are left in the library.
41% of 500 = 500 × 0.41 = 205 fiction books left,
so 261 − 205 = 56 fiction books were borrowed.
[4 marks available — 1 mark for working to find the total number
of items left in the library, 1 mark for attempting to find 41% of
the total, 1 mark for finding number of fiction books left in the
library and 1 mark for the correct final answer.]
If you made a mistake when adding up the number of items left in the
library, but did the rest of the working correctly, you'd get 3 marks out
of 4 here. Questions in your exam are marked like this — so it's super
important that you show all your working.

6   a)   8% = 0.08
      £2000 × 0.08 = £160 interest each year *[1 mark]*
      £160 × 3 = £480 interest in 3 years
      £2000 + £480 = £2480 *[1 mark]*
      [2 marks available in total — as above]

   b)   £702 = 108% *[1 mark]*
      £702 ÷ 108 = £6.50 = 1% *[1 mark]*
      £6.50 × 100 = £650 = 100%
      So he had £650 in his account at the start of the year. *[1 mark]*
      [3 marks available in total — as above]

7   A ratio of 3 : 7 means that 3 out of 10 = 30%
of the animals are cats *[1 mark]*
40% of 30% = 0.4 × 30% = 12% are black cats *[1 mark]*
100% − 30% = 70% are dogs *[1 mark]*
50% of 70% = 0.5 × 70% = 35% are black dogs *[1 mark]*
So, 35% + 12% = 47% are black animals *[1 mark]*
[5 marks available in total — as above]

Page 55: Compound Growth and Decay

1   Multiplier = 1 + 0.06 = 1.06
    After 3 years she will owe: £750 × (1.06)^3 = £893.262
    = £893.26 (to the nearest penny)
    [3 marks available — 1 mark for working out the multiplier,
    1 mark for a correct method, 1 mark for the correct answer]

2   a)   When it was first opened t = 0, so the balance would have been
       B = 5000 × 1.02^0 = 5000 × 1 = £5000 *[1 mark]*
   b)   After 7 years there would be:
       B = 5000 × 1.02^7 *[1 mark]*
        = £5743.4283...
        = £5743.43 (to the nearest penny) *[1 mark]*
       [2 marks available in total — as above]

3   10% increase = 1 + 0.1 = 1.1 *[1 mark]*
    After 5 km, the car will be travelling at 30 × 1.1^5 *[1 mark]*
    = 48.3153 = 48.3 km/h (3 s.f.) *[1 mark]*
    [3 marks available in total — as above]

Pages 56-57: Unit Conversions

1   a)   1 litre = 1000 millilitres
       7.5 × 1000 = 7500
       So 7.5 litres = 7500 ml *[1 mark]*
   b)   1 stone = 14 pounds
          0 1 2
      14)1¹6²8 *[1 mark]*
       So 168 pounds = 12 stone *[1 mark]*
       [2 marks available in total — as above]

2   870 mm *[1 mark]*

3   a)   2 m = 200 cm. $\frac{47}{200} = \frac{23.5}{100}$ = 23.5%

       [2 marks available — 1 mark for converting measurements
       to the same unit, 1 mark for the correct percentage]

   b)   3 feet = 36 inches. $\frac{9}{36} = \frac{1}{4}$

       [2 marks available — 1 mark for converting measurements
       to the same unit, 1 mark for the correct fraction]

4   64 pints = 64 ÷ 8 = 8 gallons *[1 mark]*
    8 gallons = 4 × 2 gallons ≈ 4 × 9 litres *[1 mark]*
    = 36 litres *[1 mark]*
    [3 marks available in total — as above]

5   One book weighs 0.55 lb so 8 books will weigh
    8 × 0.55 lb = 4.4 lb *[1 mark]*
    1 kg ≈ 2.2 lb
    4.4 ÷ 2.2 = 2
    So the eight books weigh 2 kg. *[1 mark]*
    1 kg = 1000 g
    2 × 1000 = 2000
    So the books weigh 2000 g. *[1 mark]*
    For 100 g postage is £0.50 so for 2000 g postage is
    £0.50 × 20 = £10. *[1 mark]*
    [4 marks available in total — as above]

6   1.6 m = 160 cm and 1.5 m = 150 cm
    160 ÷ 20 = 8, so he can fit 8 tiles along the 1.6 m side
    150 ÷ 20 = 7.5, so he can fit 7.5 tiles along the 1.5 m side
    So he needs 8 × 7.5 = 60 tiles to cover the wall exactly.
    [3 marks available — 1 mark for converting all measurements to
    the same unit, 1 mark for working out how many will fit along
    each side, 1 mark for the correct answer]

7   Start by converting the side lengths to the same measurement
    1 m = 100 cm = 1000 mm
    3 × 1000 = 3000
    So 3 m = 3000 mm *[1 mark]*
    Work out how many small cubes you could fit along each side of
    the large cube: 3000 ÷ 60 = 50 *[1 mark]*
    So in the large cube you could fit:
    50 × 50 × 50 = 125 000 small cubes *[1 mark]*
    [3 marks available in total — as above]

Page 58: Time Intervals

1   a)   She has to wait from 16 58 till 17 04 which
       is 6 minutes. *[1 mark]*
   b)   It'll take from 16 40 to 18 15:
       16 40 till 17 00 is 20 minutes.
       17 00 till 18 00 is 1 hour.
       18 00 till 18 15 is 15 minutes.
       So the total time is:
       1 hour + 20 minutes + 15 minutes = 1 hour 35 minutes
       [2 marks available in total — 1 mark for a correct method,
       1 mark for the correct answer]

2   9:55 + 2 hours = 11:55
    11:55 + 15 minutes = 12:10
    The cake is 400 g so she needs to bake it for an extra
    10 × 4 = 40 minutes
    12:10 + 40 minutes = 12:50
    [3 marks available — 1 mark for calculating that the cake must
    be baked for an extra 40 minutes, 1 mark for a correct method of
    adding times and 1 mark for the correct final answer]

3   4.30 pm till 5.00 pm is 30 minutes.
    5.00 pm till 7.00 pm is 2 hours.
    7.00 pm till 7.15 pm is 15 minutes.
    So they spend:
    2 hours + 30 minutes + 15 minutes = 2 hours 45 minutes
    2 hours 45 minutes = 2.75 hours
    2.75 × 12 = 33 hours
    33 hours + 7 hours 10 minutes = 40 hours 10 minutes
    [4 marks available — 1 mark for a correct method to find the
    time from 4.30 pm till 7.15 pm, 1 mark for finding the correct
    time from 4.30 pm till 7.15 pm, 1 mark for the correct total time
    for the first 12 days, 1 mark for the correct answer]

Page 59: Speed

1   1 hour 15 minutes = 1.25 hours *[1 mark]*
    Distance = speed × time, so distance = 56 × 1.25 = 70 km *[1 mark]*
    [2 marks available in total — as above]

2   a)   Speed = Distance ÷ Time, so speed = 36 ÷ 5
                          = 7.2 km/h *[1 mark]*
   b)   1 km = 1000 m
       7.2 × 1000 m = 7200 m
       So 7.2 km/h = 7200 m/h *[1 mark]*
       1 hour = 60 minutes = 3600 seconds
       7200 ÷ 3600 = 2 m/s
       So the average speed of the giraffe is 2 m/s *[1 mark]*
       [2 marks available in total — as above]
       You could also convert the original measurements into metres and
       seconds and then do another speed calculation.

3   a)   1 mile ≈ 1.6 km
       2.5 ÷ 1.6 = 1.5625.
       So 2.5 km = 1.5625 miles *[1 mark]*
       1 hour = 60 minutes, so 3 minutes = 3 ÷ 60
                              = 0.05 hours *[1 mark]*
       Speed = 1.5625 miles ÷ 0.05 hours
              = 31.25 mph = 31 (to nearest mph) *[1 mark]*
       [3 marks available in total — as above]
       It doesn't matter whether you do the conversion to miles per hour
       at the start or the end of the calculation — you could find the
       speed in km/minute or km/h or miles/minute, and then change it
       to mph. Whichever way, you should get the same answer.
   b)   E.g. time = 1.5625 miles ÷ 30 mph = 0.05208... hours
       0.05208... hours × 60 × 60 = 187.5 seconds
       = 188 seconds (to nearest second)
       [2 marks available — 1 mark for dividing the distance
       by the speed limit, 1 mark for the correct answer]

Page 60: Density and Pressure

1 a) Volume = 360 ÷ 1800 *[1 mark]*
       = 0.2 m³ *[1 mark]*
       [2 marks available in total — as above]
   b) Density = 220 ÷ 0.2 *[1 mark]*
       = 1100 kg/m³ *[1 mark]*
       [2 marks available in total — as above]

2 a) Volume = mass ÷ density
       Volume of metal A = 120 ÷ 6 = 20 cm³ *[1 mark]*
       Volume of metal B = 130 ÷ 5 = 26 cm³ *[1 mark]*
       Total volume = 20 + 26 = 46 cm³ *[1 mark]*
       [3 marks available in total — as above]
   b) Density = mass ÷ volume = (120 + 130) ÷ 46
       = 250 ÷ 46 *[1 mark]* = 5.43478... = 5.4 g/cm³ (1 d.p.) *[1 mark]*
       [2 marks available in total — as above]

3 Area of face A = 2 m × 4 m = 8 m² *[1 mark]*
   Pressure = Force ÷ Area = 40 N ÷ 8 m² *[1 mark]*
       = 5 N/m² *[1 mark]*
   [3 marks available in total — as above]

Section Five — Shapes and Area

Page 61: Properties of 2D Shapes

1 a) Isosceles triangle *[1 mark]*
       You need to say "isosceles triangle" to get the mark, not just "triangle".
   b) C *[1 mark]*

2 a) No lines of symmetry *[1 mark]*
   b) Order 2 *[1 mark]*

3 $x = 3$ *[1 mark]*

4 Rhombuses have two pairs of equal angles, so one of the other
   angles must be 62°. *[1 mark]*
   Neighbouring angles add up to 180°, so the other angles both equal
   180° − 62° = 118°. *[1 mark]*
   [2 marks available in total — as above]

Page 62: Congruent and Similar Shapes

1 The triangles are right-angled, so by Pythagoras' theorem,
   $BC^2 = AB^2 + AC^2 = 8^2 + 6^2 = 100$. So $BC = \sqrt{100} = 10$ cm.
   If the triangles were congruent, BC would be the same as EF
   (by RHS), but $10 \neq 11$, so the triangles are not congruent.
   *[2 marks available — 1 mark for calculating the length of BC or
   DE, 1 mark for explaining why the triangles are not congruent]*

2 a) Scale factor from $EFGH$ to $ABCD$ = 9 ÷ 6 = 1.5 *[1 mark]*
       EF = 6 ÷ 1.5 = 4 cm *[1 mark]*
       [2 marks available in total — as above]
   b) BC = 4 × 1.5 = 6 cm *[1 mark]*

3 Angle EBD = 180° − 55° − 65° = 60°
   Angle x = angle EBD = 60° (vertically opposite angles) *[1 mark]*
   Scale factor from ABC to DBE = 5 ÷ 2 = 2.5 *[1 mark]*
   So y = 6 ÷ 2.5 = 2.4 cm *[1 mark]*
   [3 marks available in total — as above]
   *To answer this question, you need to know that vertically opposite
   angles are equal — see section 6.*

Pages 63-64: The Four Transformations

1
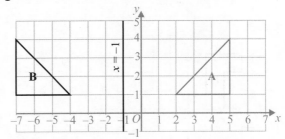

*[2 marks available — 2 marks for correct reflection,
otherwise 1 mark for triangle reflected but in wrong position]*

2 a) $\begin{pmatrix} 2 \\ -5 \end{pmatrix}$
       *[2 marks available — 1 mark for $\begin{pmatrix} \pm 2 \\ \pm 5 \end{pmatrix}$, 1 mark for fully
       correct answer]*
   b)
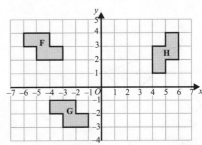

*[2 marks available — 1 mark for a rotation of 90° clockwise
about any point, 1 mark for correct centre of rotation]*

3
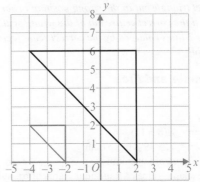

*[3 marks available — 3 marks for correct enlargement, otherwise
2 marks for a correct triangle but in the wrong position or for
an enlargement from the correct centre but of the wrong scale
factor, or 1 mark for 2 lines enlarged by the correct scale factor
anywhere on the grid]*

4 a)
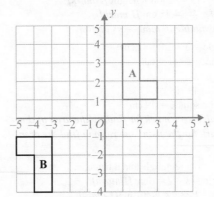

*[2 marks available — 1 mark for rotation of 180° about
any point, 1 mark for the correct centre of rotation]*

Answers

156

b)

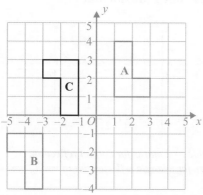

[1 mark]

c) Rotation of 180° about (0, 2).
    [3 marks available — 1 mark for rotation, 1 mark for 180°,
    1 mark for the correct centre of rotation]

5

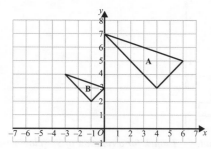

[3 marks available — 3 marks for correct enlargement, otherwise
2 marks for a correct triangle but in the wrong position or for
an enlargement from the correct centre but of the wrong scale
factor, or 1 mark for 2 lines enlarged by the correct scale factor
anywhere on the grid]

Pages 65-66: Perimeter and Area

1 a) 18 cm *[1 mark]*
   b) 14 cm² *[1 mark]*
2 a) Area of trapezium = ½(8 + 11) × 6
                         = ½ × 19 × 6 = 57 cm²
       Area of triangle = area of trapezium ÷ 3 = 57 ÷ 3 = 19 cm²
       Total area of the shape = area of trapezium + area of triangle
                               = 57 + 19 = 76 cm²
       [3 marks available — 1 mark for the area of the trapezium,
       1 mark for the area of the triangle, 1 mark for correct final
       answer]
   b) Area of triangle = ½ × base × height
                     19 = ½ × 8 × height *[1 mark]*
       height = 19 ÷ 4 = 4.75 cm *[1 mark]*
       [2 marks available in total — as above]
3 Area of rectangle = 3 × 14 = 42 cm²
   area of rectangle : area of square
                    6 : 7
        (×7)   42 : 49   (×7)
   So the area of the square = 49 cm²
   [2 marks available — 1 mark for finding the area of the
   rectangle, 1 mark for the using the ratio to find the area
   of the square]
4 Split the shape into a rectangle and a triangle.
   Area of rectangle = 3 × 2 = 6 m² *[1 mark]*
   Area of triangle = ½ × 2 × 1 = 1 m² *[1 mark]*
   Area of logo = 6 + 1 = 7 m²
   1 tin covers 3 m², and 7 ÷ 3 = 2.333... *[1 mark]*
   so she will need 3 tins of paint *[1 mark]*.
   [4 marks available in total — as above]
   You have to round up here, as two tins wouldn't be enough.

5 Area of patio = 5 × 5 = 25 m²
   Area of lawn and patio = 27 × 10 = 270 m²
   Area of lawn = 270 − 25 = 245 m² *[1 mark]*
   245 ÷ 10 = 24.5, so 25 boxes needed. *[1 mark]*
   Cost = no. of boxes × price per box = 25 × 7
        = £175.00 *[1 mark]*
   [3 marks available in total — as above]

6 Area of rectangle = 6 × 8 = 48 cm² *[1 mark]*
   Base of triangle = 8 cm − 5 cm = 3 cm
   Height of triangle = 6 cm − 2 cm = 4 cm
                              [1 mark for base and height]
   Area of triangle = $\frac{1}{2}$ × 3 × 4 = 6 cm² *[1 mark]*
   Area of shaded area = 48 − 6 = 42 cm² *[1 mark]*
   [4 marks available in total — as above]

Pages 67-68: Perimeter and Area — Circles

1 a) Diameter *[1 mark]*
   b) Radius *[1 mark]*
   c) Chord *[1 mark]*
   d)

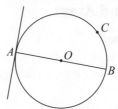

       [1 mark for line drawn at right angles to the end of the
       diameter at A]
2 a) Circumference = π × (2 × 0.25) *[1 mark]*
                     = 1.57 m (to 2 d.p.) *[1 mark]*
       [2 marks available in total — as above]
   b) 500 ÷ 1.57 = 318.47... *[1 mark]*
       So the wheel makes 318 full turns. *[1 mark]*
       [2 marks available in total — as above]
3 Area of circular card = π × 5² = 25π cm² *[1 mark]*
   Area of cut out circle = π × 3² = 9π cm² *[1 mark]*
   Area of letter "O" = 25π − 9π = 16π cm² *[1 mark]*
   [3 marks available in total — as above]
4 a) Area = π × 14² ÷ 2 *[1 mark]*
           = 307.876... = 308 mm² (3 s.f.) *[1 mark]*
       [2 marks available in total — as above]
   b) Length of curved edge = [(π × (14 × 2)) ÷ 2]
                             = 43.982...*[1 mark]*
       Perimeter = 43.982... + (14 × 2)
                 = 71.982... = 72.0 mm (3 s.f.) *[1 mark]*
       [2 marks available in total — as above]
5 Area of square = 8 × 8 = 64 m² *[1 mark]*
   Area of circle = π × 4² = 50.2654... m² *[1 mark]*
   Shaded area = 64 − 50.2654... = 13.7345... m²
   = 13.73 m² (2 d.p.) *[1 mark]*
   [3 marks available in total — as above]
6 Circumference of full circle = 2 × π × 6 = 12π cm
   Length of arc = $\frac{30}{360}$ × circumference of circle
   = $\frac{30}{360}$ × 12π = π cm
   Perimeter of sector = π + 6 + 6 = 15.1415... = 15.1 cm (3 s.f.)
   Area of full circle = π × 6² = 36π cm²
   Area of sector = $\frac{30}{360}$ × area of circle
   = $\frac{30}{360}$ × 36π = 3π cm² = 9.4247... = 9.42 cm² (3 s.f.)
   [5 marks available — 1 mark for a correct method for calculating
   the length of the arc, 1 mark for correct arc length, 1 mark for
   correct perimeter of sector, 1 mark for a correct method for
   finding the area of the sector, 1 mark for correct area of sector]

Pages 69-71: 3D Shapes

1 a)

	Triangle-based pyramid	Square-based pyramid	Pentagon-based pyramid
Number of Faces	4	**5**	6
Number of Vertices	**4**	5	**6**
Number of Edges	6	**8**	10

[2 marks available — 2 marks if all four entries are correct, otherwise 1 mark if 2 or 3 entries are correct]

b) $E = 2x$

[2 marks available — 2 marks for the correct formula, otherwise 1 mark for just '2x' with no 'E =']

c) In a pyramid with an octagonal base, $x = 8$, so
$E = 2 \times 8 = 16$ *[1 mark]*

2

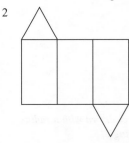

[1 mark]

3 a) Area of cross-section $\approx \frac{1}{2} \times 6 \times 5 = 15$ cm^2 *[1 mark]*
Volume of prism $\approx 15 \times 6 = 90$ cm^3 *[1 mark]*
[2 marks available in total — as above]

b) This is an overestimate, as all of the numbers have been rounded up. *[1 mark]*

4 Volume of sphere $= \frac{4}{3}\pi r^3 = \frac{4}{3} \times \pi \times 15^3$ *[1 mark]* $= 4500\pi$ cm^3
$= 14\,137.166... = 14\,100$ cm^3 (3 s.f.) *[1 mark]*
[2 marks available in total — as above]

5 a) Volume $= 90 \times 40 \times 30$ *[1 mark]* $= 108\,000$ cm^3 *[1 mark]*
[2 marks available in total — as above]

b) Volume of cuboid $=$ length $\times$ width $\times$ height
$108\,000 = 120 \times$ width $\times 18$
$108\,000 = 2160 \times$ width
width $= 108\,000 \div 2160 = 50$ cm
[2 marks available — 1 mark for correctly rearranging the formula to find the width, 1 mark for the correct answer]

6 Surface area of cube $= 6 \times$ area of one face
$= 6 \times 7 \times 7 = 294$ cm^2 *[1 mark]*
Surface area of square-based pyramid
$=$ area of base $+ (4 \times$ area of one triangular face$)$
$= 2 \times 2 + (4 \times ½ \times 2 \times 2) = 4 + 8 = 12$ cm^2 *[1 mark]*
Surface area of cube : surface area of pyramid
$= 294 : 12$ *[1 mark]*
$= 294 \div 12 : 12 \div 12 = 24.5 : 1$ *[1 mark]*
[4 marks available in total — as above]

7 Volume of water in paddling pool $= \pi \times r^2 \times h$
$= \pi \times 100^2 \times 40$ *[1 mark]* $= 400\,000\pi$ cm^3
Time it will take to fill to 40 cm $= 400\,000\pi \div 300$ *[1 mark]*
$= 4188.790...$ seconds
Convert to minutes $= 4188.790... \div 60$
$= 69.813... = 70$ minutes (to the nearest minute) *[1 mark]*
[3 marks available in total — as above]

Page 72: Projections

1

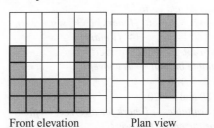

Front elevation   Plan view

[2 marks available — 1 mark for a correct front elevation diagram, 1 mark for a correct plan view diagram in any orientation]

2 Split the shape into two cuboids, looking at the front elevation. The bottom cuboid has $4 \times 2 \times 4 = 32$ cubes in it. The cuboid at the top has $2 \times 2 \times 4 = 16$ cubes in it. So there are $32 + 16 = 48$ cubes in the shape.
[2 marks available — 1 mark for a correct calculation, 1 mark for the correct answer]
You might have split your shape up differently — as long as your working is correct and you get the correct answer, you'll get all the marks.

3

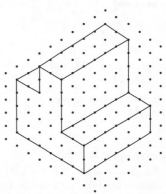

[2 marks available — 2 marks for a correct diagram, otherwise 1 mark for the correct cross-section but wrong length]

Section Six — Angles and Geometry

Pages 73-74: Five Angle Rules

1 Angles on a straight line add up to 180°,
so $x + 30° + 50° = 180°$. *[1 mark]*
$x = 180° - 50° - 30° = 100°$. *[1 mark]*
[2 marks available in total — as above]

2 $110° + 170° + 50° + 40° = 370°$. *[1 mark]*
Angles round a point add up to 360°, not 370°, so these angles do not fit round a point as they are shown on the diagram. *[1 mark]*
[2 marks available in total — as above]

3 $70° + 90° + 97° = 257°$
Angle $ADC = 360° - 257° = 103°$
(angles in a quadrilateral add up to 360°)
[2 marks available — 1 mark for a correct method, 1 mark for the correct answer]

4 Angle $CBE = 180° - 115° = 65°$
Angle $BED = 180° - 103° = 77°$ *[1 mark for both]*
$x + 90° + 77° + 65° = 360°$ *[1 mark]*
$x + 232° = 360°$
$x = 360° - 232° = 128°$ *[1 mark]*
[3 marks available in total — as above]

5 $180° - 48° = 132° = $ Angles $ACB + BAC$ *[1 mark]*
(angles in a triangle add up to 180°)
Angle $ACB = 132° ÷ 2 = 66°$ *[1 mark]* (ABC is isosceles)
Angle $BCD = 180° - 66° = 114°$ *[1 mark]*
(angles on a straight line add up to 180°)
[3 marks available in total — as above]

6 a) $2x + 3x + 4x = 180$
So $9x = 180$ and $x = 20$
*[2 marks available — 1 mark for a correct method,
1 mark for the correct answer]*

 b) $2x + y = 180$
$40 + y = 180$, so $y = 180 - 40 = 140$
*[2 marks available — 1 mark for a correct method,
1 mark for the correct answer]*

7 $3x + x + 5x + 90 + 81 = 360$
$9x + 171 = 360$ *[1 mark]*
$9x = 189$
 $x = 21$ *[1 mark]*
  Angle $ZXY = 180° - x° - 5x°$
      $= 180° - 6x°$ *[1 mark]*
      $= 180° - 126° = 54°$
so angle $ZXY = 54°$ *[1 mark]*
[4 marks available in total — as above]

Page 75: Parallel Lines

1 $a = 75°$ *[1 mark]*
because vertically opposite angles are equal. *[1 mark]*
[2 marks available in total — as above]

2 Angle $BCG = $ Angle $CGE = (5x + 10)°$ *(alternate angles)*
So $4x - 28 + 5x + 10 = 180$ *[1 mark]*
$9x = 198$ *[1 mark]*
$x = 22$ *[1 mark]*
[3 marks available in total — as above]
There are other ways to find x. For instance angles ACB and CGF
are corresponding angles. You can then use angles on a straight line
to find x.

3 Angle $FEG = 180° - 70° = 110°$
(angles on a straight line add up to 180°)
Angle $EGH = 145°$ (corresponding angles)
Angle $EGF = 180° - 145° = 35°$
(angles on a straight line add up to 180°)
Angle $EFG = 180° - 35° - 110° = 35°$
(angles in a triangle add up to 180°)
So the triangle must be isosceles as it has two equal angles.
*[3 marks available — 3 marks for proving triangle is isosceles
by showing that two of its angles are equal, otherwise 1 mark
for finding angle FEG, 1 mark for finding either angle EGF or
angle EFG]*

Page 76: Angles in Polygons

1 Size of each exterior angle of a regular pentagon:
$360° ÷ 5$ *[1 mark]* $= 72°$ *[1 mark]*
[2 marks available in total — as above]

2 Exterior angle $= 180° - 150° = 30°$ *[1 mark]*
Number of sides $= 360° ÷ 30°$ *[1 mark]*
      $= 12$ *[1 mark]*
[3 marks available in total — as above]

3 The polygon is split into 5 triangles.
Angles in a triangle add up to 180°
Angles in polygon $= 5 × 180°$
      $= 900°$
*[3 marks available — 3 marks for correct explanation, otherwise
1 mark for stating angles in triangle add up to 180° and 1 mark
for attempt at adding to find angles in polygon]*

Page 77: Triangle Construction

1
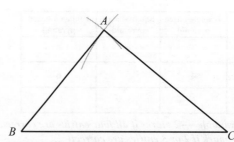
*[3 marks available — 1 mark for BC within 1 mm of 5.6 cm, 1 mark
for AB within 1 mm of 3.5 cm if correct construction arc is shown,
1 mark for AC within 1 mm of 4.3 cm if correct construction
arc is shown]*

2 a)
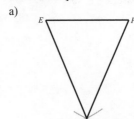
*[2 marks available — 1 mark for arcs drawn with a radius
of 4.5 cm, 1 mark for completed triangle]*

 b)
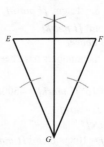
*[2 marks available — 1 mark for correct construction arcs,
1 mark for correct bisector]*

Pages 78-79: Loci and Construction

1
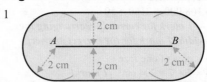
Scale: 1 cm represents 1 m
(diagram not actual size)
*[2 marks available — 2 marks for arcs with a radius of 2 cm
centred at A and B, lines parallel to AB 2 cm either side of AB
and correct area shaded, otherwise 1 mark for arcs with a radius
of 2 cm centred at A and B or for lines parallel to AB 2 cm either
side of AB]*
You'll still get the marks if you are within 1 mm of the
correct measurements.

2
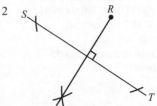
*[2 marks available — 1 marks for accurate perpendicular line
through R and 1 mark for showing all construction arcs.]*

3

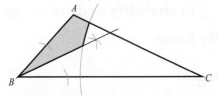

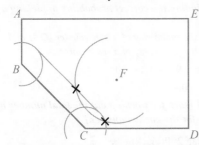

[4 marks available — 1 mark for arc with radius of 6.5 cm with centre at C, 1 mark for construction arcs on AB and BC for angle bisector at ABC, 1 mark for correct angle bisector at ABC, and 1 mark for the correct shading]
Remember to leave in your construction lines.

4

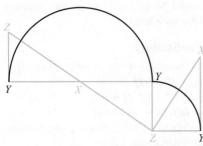

[4 marks available — 1 mark for arcs with radius of 1 cm centred at B and C, 1 mark for a line parallel to BC 1 cm from BC, 1 mark for an arc with radius of 2 cm centred at F, 1 mark for correct crosses at the intersections]

5

[3 marks available — 1 mark for constructing either the semicircle or quarter circle correctly, 1 mark for the two parts of the locus being joined together, 1 mark for a completely correct diagram]

Page 80: Bearings

1   260° (allow 258°-262°) *[1 mark]*
    It's easier to measure the 100° angle and subtract it from 360°

2   180° − 79° = 101° *(allied angles)* *[1 mark]*
    360° − 101° = 259°
    Ruth travels on a bearing of 259°. *[1 mark]*
    [2 marks available in total — as above]

3   360° − 162° − 82° = 116° *[1 mark]*
    180° − 116° = 64°
    Bearing of *B* from *A* is 064°. *[1 mark]*
    [2 marks available in total — as above]

Page 81: Maps and Scale Drawings

1   a)   Drawing of dining table is 4 cm long.
         So 4 cm is equivalent to 2 m.
         2 ÷ 4 = 0.5
         Therefore scale is 1 cm to 0.5 m *[1 mark]*

    b)   On drawing, dining table is 3 cm from shelves.
         So real distance = 3 × 0.5 = 1.5 m *[1 mark]*

    c)   The chair and the space around it would measure 4 cm × 5 cm
         on the diagram and there are no spaces that big, so no, it would
         not be possible.
         [2 marks available — 1 mark for correct answer, 1 mark for reasoning referencing diagram or size of gaps available]

2   Using the scale 1 cm = 100 m:
    400 m = 4 cm and 500 m = 5 cm

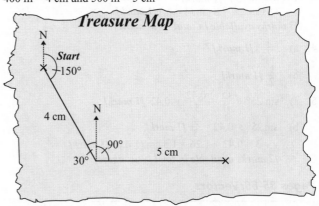

[3 marks available — 1 mark for line on accurate bearing of 150°, 1 mark for line on accurate bearing of 090°, 1 mark for accurate 4 cm and 5 cm line lengths]

Page 82: Pythagoras' Theorem

1   $AB^2 = 4^2 + 8^2$ *[1 mark]*
    $AB^2 = 16 + 64 = 80$
    $AB = \sqrt{80}$ *[1 mark]*
    $AB = 8.94$ cm (2 d.p) *[1 mark]*
    [3 marks available in total — as above]

2   The triangle can be split into two right-angled triangles.

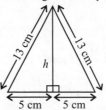

Let *h* be the height of the triangle:
$13^2 = 5^2 + h^2$ *[1 mark]*
$h^2 = 169 − 25 = 144$
$h = \sqrt{144}$ *[1 mark]*
$h = 12$ cm *[1 mark]*
[3 marks available in total — as above]

3   Let *b* be the width of the rectangle. The rectangle is split into two
    right-angled triangles with sides of length *b* cm, 3 cm and 5 cm.
    $5^2 = 3^2 + b^2$ *[1 mark]*
    $b^2 = 25 − 9 = 16$
    $b = \sqrt{16}$ *[1 mark]*
    $b = 4$ cm *[1 mark]*
    Area of rectangle = 3 cm × 4 cm
    = 12 cm² *[1 mark]*
    [4 marks available in total — as above]

Pages 83-84: Trigonometry

1   $\cos x = \dfrac{A}{H}$

   $\cos 45° = \dfrac{BC}{5}$ *[1 mark]*

   $5 \times \cos 45° = BC$ *[1 mark]*

   $BC = 3.5355... = 3.5$ cm (1 d.p) *[1 mark]*

   [3 marks available in total — as above]

2   $\sin x = \dfrac{14}{18}$ *[1 mark]*

   $x = \sin^{-1}\left(\dfrac{14}{18}\right)$ *[1 mark]*

   $x = 51.0575... = 51.1°$ (1 d.p) *[1 mark]*

   [3 marks available in total — as above]

3   $\tan 38° = \dfrac{x}{10}$ *[1 mark]*

   $x = 10 \times \tan 38°$ *[1 mark]*

   $x = 7.8128... = 7.8$ cm (1 d.p.) *[1 mark]*

   [3 marks available in total — as above]

4   a) $\dfrac{1}{\sqrt{2}}$ *[1 mark]*

   b) $\dfrac{1}{2}$ *[1 mark]*

5   a) $\sin 25° = \dfrac{O}{H} = \dfrac{0.42}{1} = 0.42$ *[1 mark]*

   b) $\sin 25° = 0.42 = \dfrac{y}{3}$ *[1 mark]*

   $y = 3 \times 0.42 = 1.26 = 1.3$ cm (1 d.p) *[1 mark]*

   [2 marks available in total — as above]

Pages 85-86: Vectors

1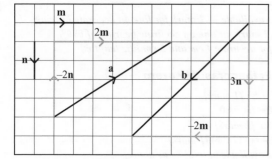

   a) $2\mathbf{m} - 2\mathbf{n}$ *[1 mark]*

   b) $3\mathbf{n} - 2\mathbf{m}$ *[1 mark]*

2   a) $\begin{pmatrix} -3 \\ 5 \end{pmatrix} - \begin{pmatrix} 5 \\ 4 \end{pmatrix} = \begin{pmatrix} -8 \\ 1 \end{pmatrix}$ *[1 mark]*

   b) $4 \times \begin{pmatrix} 5 \\ 4 \end{pmatrix} - \begin{pmatrix} -4 \\ -6 \end{pmatrix} = \begin{pmatrix} 20 \\ 16 \end{pmatrix} - \begin{pmatrix} -4 \\ -6 \end{pmatrix} = \begin{pmatrix} 24 \\ 22 \end{pmatrix}$ *[1 mark]*

   c) $2 \times \begin{pmatrix} -3 \\ 5 \end{pmatrix} + \begin{pmatrix} 5 \\ 4 \end{pmatrix} + 3 \times \begin{pmatrix} -4 \\ -6 \end{pmatrix} = \begin{pmatrix} -6 \\ 10 \end{pmatrix} + \begin{pmatrix} 5 \\ 4 \end{pmatrix} + \begin{pmatrix} -12 \\ -18 \end{pmatrix}$

   $= \begin{pmatrix} -13 \\ -4 \end{pmatrix}$ *[1 mark]*

3   a) $\overrightarrow{AC} = \overrightarrow{AB} + \overrightarrow{BC}$ *[1 mark]*

   $= 2\mathbf{c} + 2\mathbf{d}$ *[1 mark]*

   [2 marks available in total — as above]

   b) $\overrightarrow{AL} = \dfrac{1}{2} \times \overrightarrow{AC} = \dfrac{1}{2} \times 2\mathbf{c} + \dfrac{1}{2} \times 2\mathbf{d}$ *[1 mark]*

   $= \mathbf{c} + \mathbf{d}$ *[1 mark]*

   [2 marks available in total — as above]

   c) $\overrightarrow{BL} = \overrightarrow{BA} + \overrightarrow{AL}$ *[1 mark]*

   $= -2\mathbf{c} + \mathbf{c} + \mathbf{d} = -\mathbf{c} + \mathbf{d}$ *[1 mark]*

   [2 marks available in total — as above]

4   a) $\overrightarrow{BC} = -\overrightarrow{CB} = -(-6\mathbf{a} + 4\mathbf{b}) = 6\mathbf{a} - 4\mathbf{b}$

   $\overrightarrow{AC} = \overrightarrow{AB} + \overrightarrow{BC}$ *[1 mark]*

   $= 3\mathbf{a} + \mathbf{b} + 6\mathbf{a} - 4\mathbf{b} = 9\mathbf{a} - 3\mathbf{b}$ *[1 mark]*

   [2 marks available in total — as above]

   b) $\overrightarrow{AP} = \overrightarrow{AB} + \dfrac{1}{2}\overrightarrow{BC}$ *[1 mark]*

   $= 3\mathbf{a} + \mathbf{b} + \dfrac{1}{2} \times (6\mathbf{a} - 4\mathbf{b})$

   $= 6\mathbf{a} - \mathbf{b}$ *[1 mark]*

   [2 marks available in total — as above]

Section Seven — Probability and Statistics

Page 87: Probability Basics

1   a) Blue *[1 mark]*

   b) $P(\text{pink}) = \dfrac{3}{8}$ *[1 mark]*

2   $10 - 4 = 6$ red counters.

   $P(\text{red}) = \dfrac{6}{10} = 0.6$

   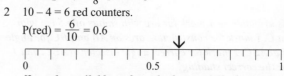

   [2 marks available — 2 marks for correctly drawn arrow, otherwise 1 mark for finding the correct probability of picking a red counter.]

   You could also work out the probability of a blue counter (0.4) and subtract it from 1 to get the probability of a red counter.

3   Total team members $= 6 + 9 + 4 + 1 = 20$

   So $P(\text{midfielder}) = \dfrac{9}{20} = 0.45$

   [2 marks available — 1 mark for working out the total number of team members, 1 mark for the correct answer.]

4   Bag 1: $P(\text{black}) = \dfrac{5}{8}$, Bag 2: $P(\text{black}) = \dfrac{2}{4} = \dfrac{4}{8}$

   Bag 3: $P(\text{black}) = \dfrac{3}{8}$

   Bag 1 would give the greatest chance of winning as the probability of picking a black ball is greatest for that bag.

   [2 marks available — 1 mark for saying bag 1 gives the greatest chance of winning, 1 mark for a correct explanation involving working out P(black) for each bag.]

Pages 88-89: More Probability

1   a) $P(\text{not spotty}) = 1 - P(\text{spotty}) = 1 - 0.25 = 0.75$ *[1 mark]*

   b) $3x = 0.75$ so $x = 0.25$ *[1 mark]*

   $P(\text{stripy}) = 2x = 0.25 \times 2 = 0.5$ *[1 mark]*

   [2 marks available in total — as above]

2   a) (Hockey, Netball), (Hockey, Choir), (Hockey, Orienteering), (Orchestra, Netball), (Orchestra, Choir), (Orchestra, Orienteering), (Drama, Netball), (Drama, Choir), (Drama, Orienteering).

   [2 marks available — 2 marks for listing all 9 correct combinations, otherwise 1 mark if at least 5 combinations are correct.]

   b) There are 9 combinations and 1 of them is hockey and netball, so $P(\text{hockey and netball}) = \dfrac{1}{9}$ *[1 mark]*

   c) There are 9 combinations and 3 of them involve drama on Monday, so $P(\text{drama on Monday}) = \dfrac{3}{9} = \dfrac{1}{3}$ *[1 mark]*

   You could also count the choices for Monday — there are 3, and 1 of them is drama.

3   a) EHM, EMH, HME, HEM, MEH, MHE

   [2 marks available — 2 marks for listing all 6 correct combinations, otherwise 1 mark if at least 3 combinations are correct.]

   b) There are 6 possible combinations and in 3 of them she does Maths before English (HME, MEH, MHE).

   So $P(\text{Maths before English}) = \dfrac{3}{6} = \dfrac{1}{2}$ *[1 mark]*

4 a)

	2	4	6	8	10
1	3	5	7	9	11
2	4	6	8	10	12
3	5	7	9	11	13
4	6	8	10	12	14
5	7	9	11	13	15
6	8	10	12	14	16

[2 marks available — 2 marks if all entries are correct, otherwise 1 mark if at least 4 entries are correct.]

b) There are 30 possible outcomes and 9 of them will score 12 or more. So $P(12 \text{ or more}) = \frac{9}{30} = \frac{3}{10}$

[2 marks available — 1 mark for finding that 9 outcomes score 12 or more, 1 mark for the correct answer.]

c) No, he is not correct. Half of the total scores are even and half are odd so he is equally likely to get an even-numbered score or an odd-numbered score.

[2 marks available — 1 mark for saying that he is not correct, 1 mark for a correct explanation.]

Pages 90-91: Probability Experiments

1 $200 \times 0.64 = 128$ times *[1 mark]*

2 $P(\text{lands on } 5) = 1 - 0.3 - 0.15 - 0.2 - 0.25 = 0.1$ *[1 mark]*
Estimate of number of times spinner lands on 5
$= 100 \times 0.1 = 10$ times *[1 mark]*
[2 marks available in total — as above]

3 a)

Number	1	2	3	4	5	6
Frequency	16	6	12	7	3	6
Relative frequency	0.32	0.12	0.24	0.14	0.06	0.12

[2 marks available — 2 marks for a fully correct table, otherwise 1 mark if at least 3 of the relative frequencies are correct.]

b) E.g. This would only be correct if the dice is fair and all outcomes are equally likely. From the table, it looks like rolling a 1 is more likely than some of the other numbers.
[2 marks available — 1 mark for saying that the dice might not be fair or that the outcomes might not be equally likely, 1 mark for using evidence from the table to explain why the dice might not be fair.]

c) The second set will give more reliable estimates as there were a greater number of trials in this experiment. *[1 mark]*

4 a)

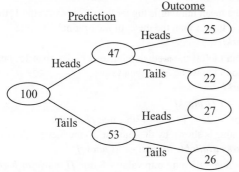

[2 marks available — 1 mark for the correct numbers for the predictions, 1 mark for the correct numbers for the outcomes.]

b) She predicted the flip correctly $25 + 26 = 51$ times out of 100 *[1 mark]*, so relative frequency $= \frac{51}{100} = 0.51$ *[1 mark]*
[2 marks available in total — as above]

5 a) Relative frequency of hitting the target with a left-handed throw $= \frac{12}{20} = \frac{3}{5}$ or 0.6.
[2 marks available — 1 mark for a correct method, 1 mark for the correct answer.]

b) E.g. The estimated probability is more reliable for his right hand because he threw the ball more times with that hand.
[1 mark]

Page 92: The AND/OR Rules

1 a) $P(4 \text{ or } 5) = P(4) + P(5)$
$= 0.25 + 0.1$ *[1 mark]*
$= 0.35$ *[1 mark]*
[2 marks available in total — as above]

b) $P(1 \text{ and } 3) = P(1) \times P(3)$
$= 0.3 \times 0.2$ *[1 mark]*
$= 0.06$ *[1 mark]*
[2 marks available in total — as above]

2 a) $P(\text{no prize}) = 1 - 0.3 = 0.7$ *[1 mark]*

b) $P(\text{no prize on either game}) = P(\text{no prize}) \times P(\text{no prize})$
$= 0.7 \times 0.7$ *[1 mark]*
$= 0.49$ *[1 mark]*
[2 marks available in total — as above]

3 $P(\text{at least 1 is late}) = 1 - P(\text{neither is late})$
$P(\text{Alisha isn't late}) = 1 - 0.9 = 0.1$ *[1 mark]*
$P(\text{Anton isn't late}) = 1 - 0.8 = 0.2$ *[1 mark]*
$P(\text{neither is late}) = 0.1 \times 0.2 = 0.02$ *[1 mark]*
$P(\text{at least 1 is late}) = 1 - 0.02 = 0.98$ *[1 mark]*
[4 marks available in total — as above]
You could also solve this question by finding P(exactly 1 is late) and P(both are late) and adding them together:
(0.1 × 0.8) + (0.9 × 0.2) + (0.8 × 0.9) = 0.98.

Page 93: Tree Diagrams

1 a)

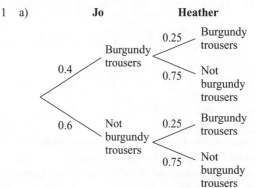

[2 marks available — 1 mark for the correct probability for Jo, 1 mark for the correct probabilities for Heather.]

b) $P(\text{neither wear burgundy trousers}) = 0.6 \times 0.75$ *[1 mark]*
$= 0.45$ *[1 mark]*
[2 marks available in total — as above]

2 a)

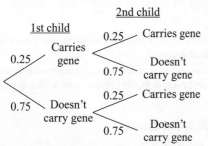

1st child — *2nd child*

Carries gene 0.25 — Carries gene
0.25 — 0.75 — Doesn't carry gene
0.75 — Doesn't carry gene — 0.25 — Carries gene
0.75 — Doesn't carry gene

[2 marks available — 1 mark for correct probabilities for 1st child, 1 mark for correct probabilities for 2nd child]

b) P(both children carry the gene) = 0.25 × 0.25 *[1 mark]*
= 0.0625 *[1 mark]*
[2 marks available in total — as above]

Page 94: Sets and Venn Diagrams

1 Number of elements in B and in A = 45 − 21 = 24
Number of elements in A but not in B = 39 − 24 = 15
Number of elements not in A or B = 60 − 24 − 21 − 15 = 0

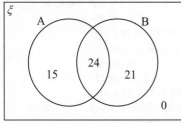

[3 marks available — 3 marks for a fully correct Venn diagram, otherwise lose 1 mark for each incorrect value.]

2 a) Number of students who only like apples = 70 − 20 = 50
Number of students who only like bananas = 40 − 20 = 20
Number of students who don't like either
= 100 − 50 − 20 − 20 = 10

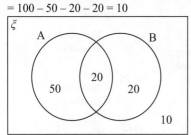

[3 marks available — 3 marks for a fully correct Venn diagram, lose 1 mark for each incorrect value.]

b) 50 + 20 + 20 = 90 students out of 100 like apples or bananas,
so P(A ∪ B) = $\frac{90}{100} = \frac{9}{10}$ or 0.9.
[2 marks available — 1 mark for 90 students liking either fruit, 1 mark for the correct answer.]

Page 95: Sampling and Data Collection

1 a) E.g.

Number of chocolate bars	Tally	Frequency
0-2		
3-5		
6-8		
9-11		
12 or more		

[2 marks available — 1 mark for a suitable tally table, 1 mark for non-overlapping classes that cover all possible values]

b) E.g. Faye's results are likely to be unrepresentative because she hasn't selected her sample at random from all the teenagers in the UK. Also, her sample is too small to represent the whole population. So Faye can't use her results to draw conclusions about teenagers in the UK.
[2 marks available — 1 mark for a correct comment based on sample size, 1 mark for stating that Faye can't draw conclusions about teenagers in the UK with reasoning]

2 a) Proportion of people in sample who travelled by car
= 22 ÷ 50 = 0.44 *[1 mark]*
Estimate of number of people at match who travelled by car
= 0.44 × 5000 *[1 mark]* = 2200 *[1 mark]*
[3 marks available in total — as above]

b) E.g. The estimate is based on a sample of 50 people, which should be big enough to give a reliable estimate.
OR E.g. A sample of 50 is too small to represent a population of 5000, so the estimate could be unreliable.
OR E.g. If the people asked weren't selected at random, then the sample might not represent the whole crowd and the estimate could be unreliable.
[1 mark for a correct comment]

Pages 96-97: Mean, Median, Mode and Range

1 a) In ascending order: 3, 10, 12, 12, 13, 18, 25, 33, 37, 41
Median = (10 + 1) ÷ 2 = 5.5th value *[1 mark]*
= (13 + 18) ÷ 2 = 15.5 minutes *[1 mark]*
[2 marks available in total — as above]

b) Mean = (3 + 10 + 12 + 12 + 13 + 18 + 25 + 33 + 37 + 41) ÷ 10
= 204 ÷ 10 *[1 mark]* = 20.4 minutes
= 20 minutes (to the nearest minute) *[1 mark]*
[2 marks available in total — as above]

c) Range = 41 − 3 = 38 minutes *[1 mark]*

2 1, 4, 7
[2 marks available — 2 marks for all three numbers correct, otherwise 1 mark for 3 numbers that have a range of 6 and a mean of 4 but aren't all different, or 3 different numbers that add up to 12 or that have a range of 6]

3 a) 23, 26, 36 (in any order)
range = 13, median = 26
[2 marks available — 1 mark for all three weights correct, 1 mark for both range and median correct]

b) Total weight of all 6 goats
= 32 + 23 + 31 + 28 + 36 + 26 = 176 kg
Weight of 4 remaining goats = 4 × 27.25 = 109 kg *[1 mark]*
176 kg − 109 kg = 67 kg *[1 mark]*
so, goats weighing 31 kg and 36 kg *[1 mark]*
[3 marks available in total — as above]

4 a) Yes, the mean number is higher than 17 because the 11th data value is higher than the mean of the original 10 values.
[1 mark]

b) You can't tell if the median number is higher than 15, because you don't know the other data values.
[1 mark]

5 a) Mode = 1 *[1 mark]*

b) Median = (25 + 1) ÷ 2 = 13th value *[1 mark]*.
13th value is shown by the 2nd bar, so median = 1 *[1 mark]*.
[2 marks available in total — as above]

6 a) Max value = 63 mm, min value = 8 mm *[1 mark for both]*,
so range = 63 − 8 = 55 mm *[1 mark]*
E.g. a range of 55 mm isn't a good reflection of the spread of the data because most of the data is much closer together.
[1 mark for a correct comment]
[3 marks available in total — as above]
You could also say that the single value of 63 mm has a big effect on increasing the value of the range so that it doesn't represent the spread of the rest of the data.

b) Median rainfall in June = (12 + 1) ÷ 2 = 6.5th value
= (29 + 30) ÷ 2 = 29.5 mm
E.g. The rainfall was generally higher in June, as the median was higher. The rainfall in June was much more varied than in November as the range was much bigger.
[3 marks available — 1 mark for calculating the median rainfall in June, 1 mark for a correct statement comparing the medians and 1 mark for a correct statement comparing the ranges]

Pages 98-100: Simple Charts and Graphs

1 a) 70 – 65 = 5 *[1 mark]*

b) Hot chocolate *[1 mark]*

c) Saturday = 50 + 40 + 35 + 20 = 145 cups
Sunday = 65 + 70 +10 + 5 = 150 cups
So more hot drinks were sold on Sunday.
[2 marks available — 1 mark for finding the correct total for either day, 1 mark for the correct answer]

d) 20 + 5 = 25 cups of herbal tea were sold in total,
and $\frac{20}{25} = \frac{4}{5}$ of them were sold on Saturday *[1 mark]*.

2 E.g.

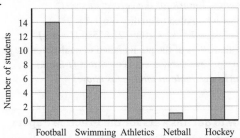

[4 marks available — 1 mark for a suitable scale starting from zero on the vertical axis, 1 mark for correctly labelling the vertical axis, 1 mark for bars of equal width and all bar heights correct, 1 mark for all bars correctly labelled]

3 a) 3 days *[1 mark]*

b) They sold more than 30 newspapers on 3 days and there are 30 days in total. So the fraction is $\frac{3}{30} = \frac{1}{10}$. *[1 mark]*

4 a) (8 × 3) + 6 = 30 *[1 mark]*

b) 40 – 20 = 20 *[1 mark]*

c)

Monday	○ ○ ○ ◔
Tuesday	○ ○ ◖
Wednesday	○ ○ ○ ○ ○
Thursday	○ ○ ○
Friday	○ ○ ○ ○ ○

[1 mark]

d) Number of eggs laid on Friday = 5 × 8 = 40 *[1 mark]*
0.4 × 40 = 16 eggs *[1 mark]*
[2 marks available in total — as above]

5 There are 4 + 2.5 + 3.5 = 10 symbols in total *[1 mark]*
So 1 symbol represents 100 ÷ 10 = 10 jars of jam *[1 mark]*
3.5 × 10 *[1 mark]* = 35 jars of raspberry jam *[1 mark]*
[4 marks available in total — as above]

6 a) 15 : 20 = 3 : 4 *[1 mark]*

b) $\frac{20}{50} \times 100$ *[1 mark]* = 40% *[1 mark]*
[2 marks available in total — as above]

7 a) The vertical axis has inconsistent numbering *[1 mark]*

b) E.g. There is a pattern that repeats every 4 points. Numbers are lowest in January, then peak in April, before decreasing in July and again in October.
[1 mark for a correct description]

Page 101: Pie Charts

1 a) $\frac{1}{4}$ *[1 mark]*

b) Badminton = 360° – 180° – 90° – 30° = 60°
Football = 180°, so 60 people = 180°
1 person = 180° ÷ 60 = 3°
So number of people who prefer badminton = 60° ÷ 3° = 20
[2 marks available — 1 mark for a correct method, 1 mark for the correct final answer]
There are other ways to work this out — e.g. you could also use the number of people who prefer football to work out the total people surveyed (120) and then multiply by the fraction who prefer badminton $\left(\frac{1}{6}\right)$.

2 a) Total number of people = 12 + 18 + 9 + 21 = 60
Multiplier = 360 ÷ 60 = 6
Plain: 12 × 6 = 72°
Salted: 18 × 6 = 108°
Sugared: 9 × 6 = 54°
Toffee: 21 × 6 = 126°

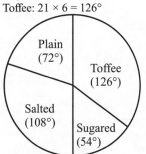

[4 marks available — 1 mark for one sector correctly drawn, 1 mark for a second sector correctly drawn, 1 mark for a complete pie chart with all angles correct, 1 mark for correct labels]

b) E.g. Chris is not right because there is no information about the number of people in the ice-cream survey. *[1 mark]*

Pages 102-103: Scatter Graphs

1 a)

[1 mark]

b) Strong positive correlation *[1 mark]*

c) E.g.

[1 mark for line of best fit passing between (10, 16) & (10, 28) and (80, 82) & (80, 96)]

2 a)

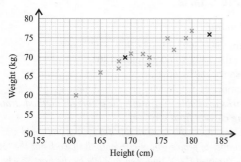

[1 mark for both points plotted correctly]

b) $\frac{5}{14}$ *[1 mark]*

c) E.g. In general, as the height increases,
the weight also increases.
[1 mark for any answer indicating a positive correlation]

3 a) E.g.

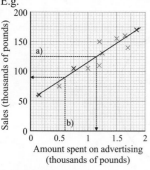

See graph — £1150
*[2 marks available — 1 mark for drawing a line of best fit,
1 mark for reading off the correct answer, allow answers
± £100]*

b) See graph above — £90 000
[1 mark, allow answers ± £10 000]

c) E.g. The estimates in a) and b) should be reliable because they
are within the range of the data.
[1 mark for a correct comment]

d) This prediction might not be reliable because £3000 is outside
the range of data and they don't know whether the same
pattern would continue.
*[1 mark for a correct comment about predicting outside the
range of the data]*

e) E.g. No, the graph doesn't prove this statement. Sales
increases might have been caused by other factors.
*[2 marks available in total — 1 mark for 'no', 1 mark for a
correct explanation]*

Page 104: Grouped Frequency Tables

1 a) $3 \le x \le 5$ *[1 mark]*

b) $(10 + 1) \div 2 = 5.5$, so the median is halfway between the 5th
and 6th values, so it lies in the group containing the 5th and 6th
values, which is $3 \le x \le 5$ *[1 mark]*

c) E.g. The mean height can't be 12 cm, because all of the heights
are less than 12 cm.
[1 mark for a correct explanation]

2 a)

Time (t secs)	Freq	Mid-interval value	Freq × Mid-interval
$22 < t \le 26$	4	$(22 + 26) \div 2 = 24$	$4 \times 24 = 96$
$26 < t \le 30$	8	$(26 + 30) \div 2 = 28$	$8 \times 28 = 224$
$30 < t \le 34$	13	$(30 + 34) \div 2 = 32$	$13 \times 32 = 416$
$34 < t \le 38$	6	$(34 + 38) \div 2 = 36$	$6 \times 36 = 216$
$38 < t \le 42$	1	$(38 + 42) \div 2 = 40$	$1 \times 40 = 40$
Total	32	—	992

Estimate of mean = 992 ÷ 32 = 31 seconds
*[4 marks available in total — 1 mark for all mid-interval
values, 1 mark for 992, 1 mark for dividing total time by total
frequency, 1 mark for the correct answer]*

b) Number who failed to qualify = 6 + 1 = 7
7 out of 32 = 7 ÷ 32 × 100 = 21.875% *[1 mark]*
More than 20% of the pupils failed to qualify for the next
round, so Anya's statement is incorrect *[1 mark]*
[2 marks available in total — as above]

How to get answers for the Practice Papers
You can download or print out worked solutions
to Practice Papers 1, 2 & 3 by going to
www.cgpbooks.co.uk/gcsemathsanswers